AF540846

GRAPH THEORY

GRAPH THEORY

By

S.C. Sharma

DISCOVERY PUBLISHING HOUSE
NEW DELHI-110002

First Published - 2007

Reprinted - 2023

ISBN: 978-81-8356-228-7

© Author

Graph Theory

Published by:

DISCOVERY PUBLISHING HOUSE

4383/4B, Ansari Road, Darya Ganj

New Delhi-110 002 (India)

Phone: +91-11-23279245; 23253475; 43596065

+91 9811179893 / +91 9871656464

E-mail: discoverybooksindia@gmail.com

orderdphbooks@gmail.com

namitwasan9@gmail.com

web: www.discoverypublishinggroup.com

Printed at:

Infinity Imaging Systems

Delhi

Preface

This book in "Graph Theory" is intended the typical-freshman or sophomore in computer science and related disciplines with a first exposure to the mathematical topics essential to their study of computer science or digital logic. There is an introduction of graph and tree in this book. This focus on the description of basic problems involving graph and tree and their applications. A major feature of this text is its versatility. Sufficient topics have been included to accommodate students with varied backgrounds.

The author will feel amply rewarded if the book serves the purpose for which it is meant. Suggestions for the improvement of the book are always welcome.

Author

CONTENTS

1

GRAPH THEORY

GRAPHS—A GENERAL INTRODUCTION

Graph theory begins with very simple geometric ideas and has many powerful applications. Because of its inherent simplicity, graph theory has a very wide range of applications in engineering, in computer sciences, in physical, social, and biological science, in linguistics and in numerous other areas. A graph can be used to represent almost any physical situation involving discrete objects and a relationship among them.

Definition 1: Directed Graph. *A directed graph consists of a set of vertices, V, and a Set of edges E, connected certain elements of V. Each element of E is an ordered pair (i.e., an element of × V). The first entry is the initial vertex of the edge and the second entry is the terminal vertex. In certain cases, there will be more than one edge between two vertices, in which case the different edges are identified with labels.*

Despite the set terminology in this definition, we usually think of a graphs a picture, an aid in visualizing a situation. We introduced this concept to help understand relations on sets. Although, those relations were principally of a mathematical nature, it remains true that when we see a graph, it tells us how the elements of a set are related to one another.

Definition 2: Simple Graph and Multigraph. *A simple graph is one for which there is no more than one edge directed from any one vertex to any other vertex. All other graphs are called multigraphs.*

To illustrate the points that we will make in this chapter, we will introduce the following examples of graphs.

Example 1:

A Directed Graph: An example of a simple directed graph. In set terms, this graph is (V, E), where V = {s, a, b} and E = {(s, a), (s, b), (a, b), (b, a), (b, b)}. Note how each edge is labeled either 0 or 1. There are often reasons for labeling even simple graphs. Some labels are to help make a graph easier to discuss; others are more significant. We will discuss the significance of the labels on this graph later.

Example 2:

An Undirected Graph. A network of computers can be easily described using a graph. FA network of five computers, a, b, c, d and e. An edge between any two vertices indicates that direct two-way communication is possible between the two computers. Note that the edges of this graph are not directed. This is due to the fact that the relation that is being displayed is symmetric (i.e., if X can communicate with Y, then Y can communicate with X). Although, directed edges could be used here, it would simply clutter the graph.

There are several other situations for which this graph can serve as a model. One of them is to interpret the vertices as cities and the edges as roads, an abstraction of a map such as the one. Another interpretation is as an abstraction of the floor plan of a house. Vertex a represents the outside of the house; all others represent rooms. Two vertices are connected if there is a door between them.

TRAVERSE—EULERIAN AND HAMILTONIAN GRAPHS

The subject of graph traversals has a long history. In fact, the solution by Leonhard Euler (Switzerland, 1707-83) of the Konigsberg Bridge Problem is considered by many to represent the birth of graph theory.

THE KONIGSBERG BRIDGE PROBLEM AND EULERIAN GRAPHS

A map of the Prussian city of Koingsberg (circa 1735) that there were seven bridges connecting the four land masses that made up the city. The legend of this problem states that the citizens of Konigsberg searched in vain for a walking tour that passed over each bridge exactly once. No one could design such a tour and the search was abruptly abandoned with the publication of Euler's Theorem.

Example 1:

(a) The degrees of A, B, C and D in 3, 3, 5 and 3, respectively.

(b) In a tournament graph, outeg (v) is the number of wins for v and indeg(v) is the number of losses. In a complete (round-robin) tournament graph with n vertices, outdeg (v) + indeg(v) = n – 1 for each vertex.

Example 2:

The complete undirected graphs K_2 and K_{2i+1}, i = 1, 2, 3, ... are Eluerian. If i is greater than one, then K_{2i} is not Eulerian.

GRAPHS

A *graph* or a *linear graph* G = (V, E) consists of a set of objects V = {b_1, b_2, ----} called *vertices or nodes* and another set E = {e_1, e_2, ---} whose elements are called *edges,* such that each edge e_k is identified with an unordered pair (v_i, v_j) of vertices. The vertices v_i, v_j associated with edge e_k are called the *end vertices* of e_k. The most common representation of a graph is by means of a diagram, in which the vertices are represented as points and each edge as a line segment joining its end vertices.

If an edge $e \in E$ is associated with an ordered pair (u, v) or an unordered pair (u, v), where $u, v \in V$, then e is said to *connect* u and v and u and v are called end points of e. An edge is said to be *incident* with the vertices it joins. Thus, the edge e that joints the nodes u and v is said to be *incident* on each of its end points u and v. Any pair of nodes that is connected by an edge in a graph is called *adjacent nodes.*

In a graph a node that is not adjacent to another node is called an *isolated node.*

Incidence and Degree

When a vertex v_i is an end vertex of some edge e_j, b_i and e_j are said to be *incident* with each other. In edge e_2, e_6 and e_7 are incident with vertex b_4. Two non-parallel edges are said to be *adjacent* if they are incident on a common vertex. For example, e_2 and e_7 are adjacent. Again, two vertices are said to be adjacent if they are the end vertices of the some edges v_4 and v_5 are adjacent, but v_1 and v_4 are not.

The number of edges incident on a vertex v_i, with self-loops counted twice, is called the *degree* of vertex v_i and is denoted by d (v_i).

$$d(v_1) = d(v_3) = d(v_4)$$
$$= 3,\ d(v_2) = 4\ ,$$

and $d(v_5) = 1$.

The degree of a vertex is also called its *valency*. Then

$d(v_1) + d(v_2) + d(v_3) + d(v_4) + d(v_5) = 3 + 4 + 3 + 3 + 1 = 14$ = twice the number of edges.

Then, we can write

$$\sum_{i=1}^{n} d(v_i) = 2e \qquad \text{....(1)}$$

Theorem:

The number of vertices of odd degree in a graph is always even.

If we consider the vertices with odd and even degrees seperately, then

$$\sum_{i=1}^{n} d(v_i) = \sum_{\text{even}} d(v_j) + \sum_{\text{odd}} d(v_k).$$

Since, by eqn. (1) the left hand side of the above equation is even, and the first expression on the right hand side is even (being a sum of even numbers), the secend expression must also be even :

$$\sum_{\text{odd}} d(v_k). \quad = \text{an even number} \qquad \text{....(2)}$$

Because in eqn. (2) each d (v_k) is odd, the total number of terms in the sum must be even to make the sum an even number.

Isolated Vertex, Pendant Vertex and Null Graph

A vertex having no incident edge is called an isolated vertex, i.e., isolated vertices are vertices with zero degree. In Fig. 1.1, vertices v_5 and v_7 are isolated vertices.

A vertex of degree one is called *pendant vertex* or an *end* vertex. In Fig 1.1, v_3 is a pendant vertex. Two adjacent edges are said to be in *series* if their common vertex is of degree two. The two edges incident on v_1, in the figure, are in series.

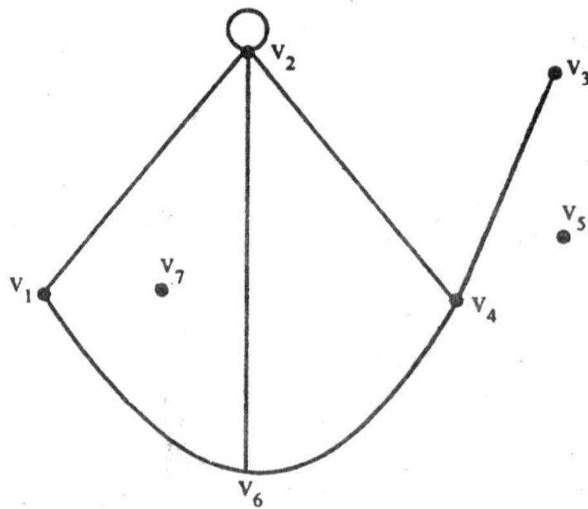

Fig. 1.1

In a graph, it is possible for the edges set E to be empty. Such a graph, without any edges is called a *null graph* obviously, every vertex in a null graph is an isolated vertex. A null graph of four vertices is shown in Fig. 1.2

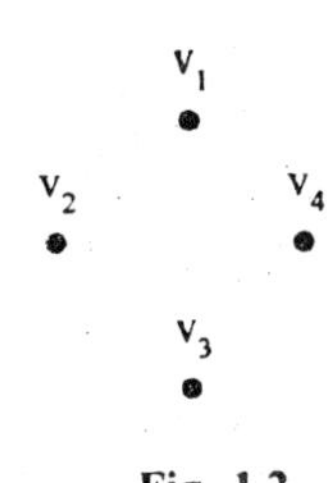

Fig. 1.2

Although the edge set E may be empty, the vertex set V must not be empty; otherwise, there will be no graph. Thus, *a graph must have at least one vertex.*

HAMILTONIAN GRAPHS

To search for a path that uses every vertex of a graph exactly once seems to be a natural next problem after you have considered Eulerian graphs. The Irish mathematician Sir William Hamilton (1805-65) is given credit for first defining such paths.

Example 1:

Agraph that is Hamiltonian. In fact it is the graph that Hamilton used as an example to pose the question of existence of Hamiltonian paths in 1859. In its original form, the puzzle that was posed to readers was called "Around the World". The vertices were labeled with names of major cities of the world and the object was to complete a tour of these cities.

Unfortunately, a simple condition doesn't exist that characterizes a Hamiltonian graph. An obvious necessary condition is that the graph be connected; however, there is a connected undirected graph with four vertices that is not Hamiltonian. Can you draw such a graph?

GRAPH OPTIMIZATION

The common thread that connects all of the problems in this section is the desire to optimize (maximize or minimize) a quantity that is associated with a graph. We will concentrate most of our attention on two of these problems, the Travelling Salesman Problem and Maximum Flow Problem. At the close of this section, we will discuss some other common optimization problems.

Definition: Weighted Graph. *A weighted graph (V, E, w), is a graph (V, E) together with a weight function* $w: E \rightarrow \mathbf{R}$. *If* $e \in E$, *w (e) is the weight on edge e.*

As you will see in our example, w (e) is usually a cost associated with the edge e; therefore, most weights will be positive.

Example 1:

Let V be the set of six capital cities in New England: Boston, Augusta, Hartford, Providence, Concord, and Montpelier. Let E be {{a, b} ∈ V : a ≠ b}; i.e., (V, E) is a complete unordered graph. A weight function on this graph is w (c_1, c_2) = the distance from c_1 to c_2. Many road maps define distance functions.

TYPES OF GRAPHS

Complete Graph

A simple graph G is said to be complete if every vertex in G is connected with every other vertex, i.e., if G contains exactly are edge between each pair of distinct vertices. A complete graph may be denoted by K_n. Some complete graphs are shown in Fig. 1.3.

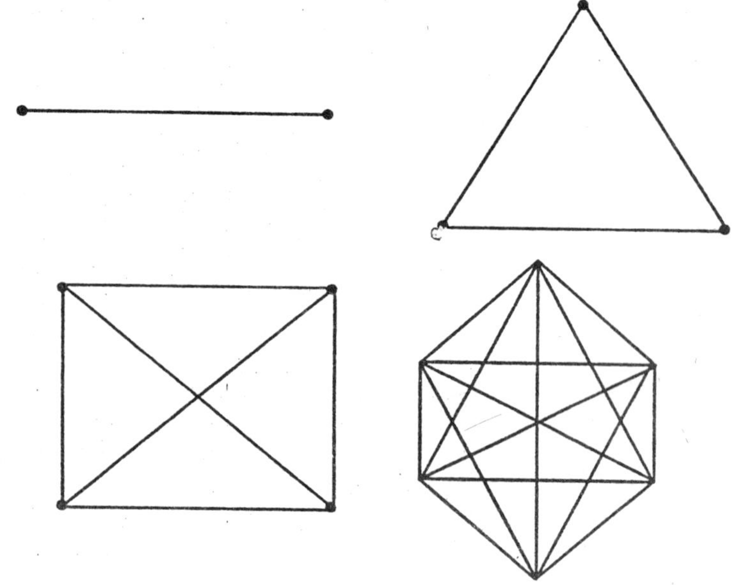

Fig. 1.3

Regular Graph

A graph in which all vertices are of equal degree is called a regular graph. If the degree of each vertex is k then the graph is called a *regular graph of degree k.*

Every null graph is regular graph of degree zero, and that the complete graph k_n is a regular graph of degree n – 1. Also, if G has n vertices and is regular of degree k, then G has 1/2 k n edges.

Platonic Graph

The graphs formed by the vertices and edges of the five regular (Platonic) solids–the tetrahedron, octahedran, cube, dodecahedron and icosahedron are known as *platononic graphs*. Some of these are shown in Fig. 1.4

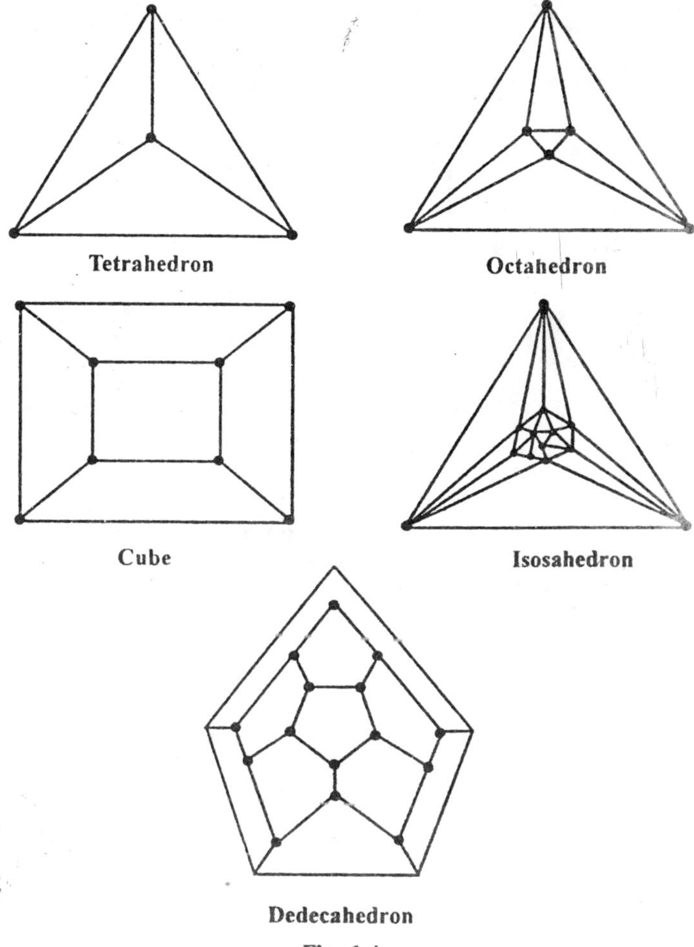

Dedecahedron

Fig. 1.4

Wheels

The wheel W_n is obtained when we have an additional vertex to the cycle C_n, for $n \geq 3$, and we connect this new vertex to each of the n vertices in C_n, by new edges. The wheels W_3, W_4, W_5 and W_6 are shown in Fig. 1.5.

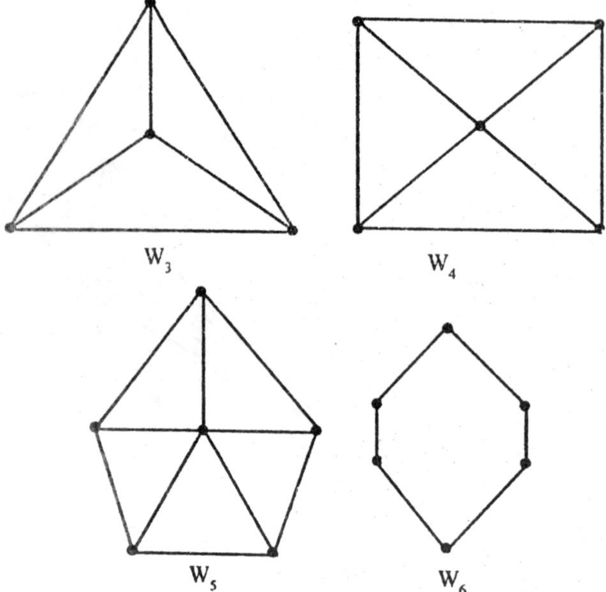

Fig. 1.5

N – Cube : The N–cube denoted by Q_n, is the graph that has vertices representing 2^n bit strings of length n. Two vertices are adjacent if and only if the bit strings that they represent differ in exactly one bit position. The graphs Q_2, Q_3 are shown in Fig 1.6. Q_n has 2^n vertices and n. 2^{n-1} edges, and is regular of degree n.

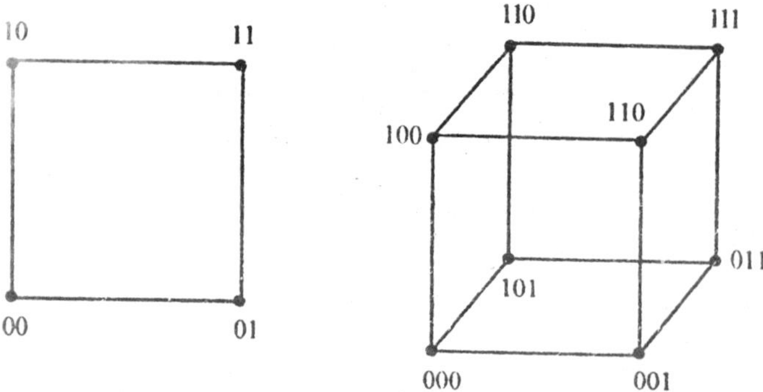

Fig. 1.6

Bipartite Graph

A graph G = (V, E) is bipartite if the vertex set V can be partitioned in two subsets V_1 and V_2 such that every edge in E connects a vertex in V_1 and a vertex in V_2. A bipartite graph can have no loop. Fig. 1.7 shows a bipartite graph.

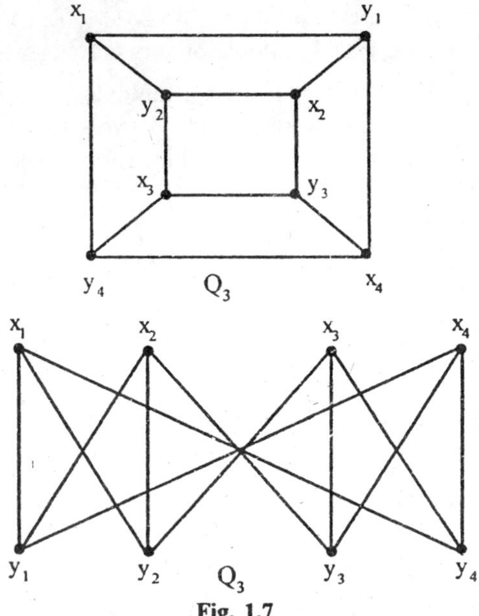

Fig. 1.7

Complete Bipartite Graph

The complete bipartite graph on m and n vertices, denoted by $K_{m,n}$ is the graph, whose vertex set is partitioned into sets V_1 with m vertices and v_2 with n vertices in which there is an edge between each pair of vertices v_1 and v_2, where $v_1 \in V_1$ and $v_2 \in V_2$. The complete bipartite graphs $K_{2,4}$ and $K_{3,3}$ are shown in Fig. 1.8. Obviously, $K_{r,s}$ has r + s vertices and r s edges.

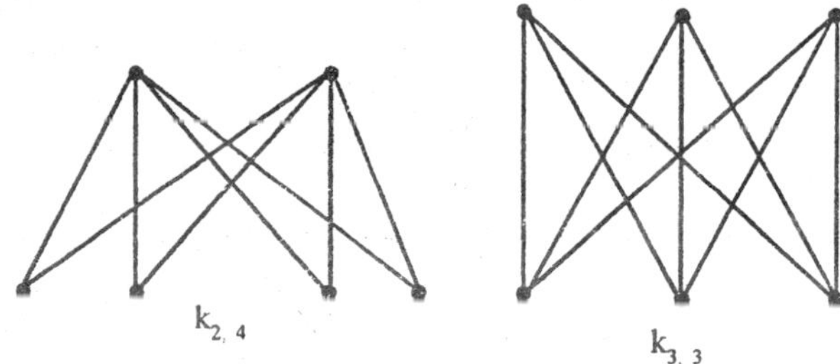

Fig. 1.8

SIMPLE GRAPH, MULTIGRAPH AND PSEUDOGRAPH

An edge of a graph that joins a node to itself is called a *loop* or *self-loop*. A loop is an edge (v_i, v_j) where $v_i = v_j$.

In some graphs, we may have pair of nodes joined by more than one edges, such edges are called *multiple* or *parellel* edges.

A graph which has neither loops nor multiple edges, i.e., where each edges connects two distinct vertices and no two edges connect the same pair of vertices is called a *simple graph*. Fig. 1.9 represents simple graphs.

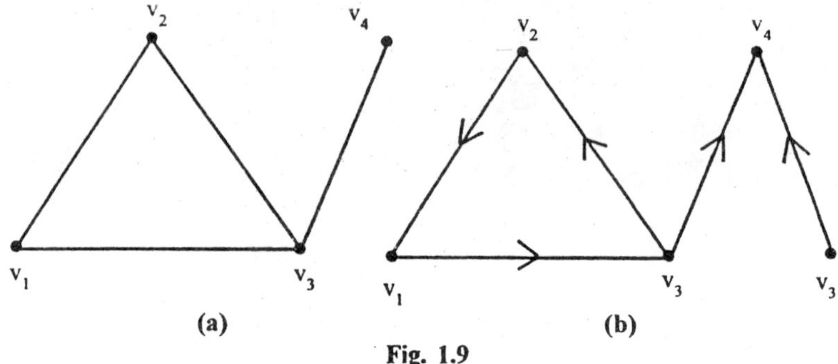

Fig. 1.9

Any graph which contains some multiple edges is called a *multigraph*. In a multigraph, no loops are allowed. Fig 1.10(a) shows a undirected multigraph while Fig. 1.10(b) represents a directed multigraph.

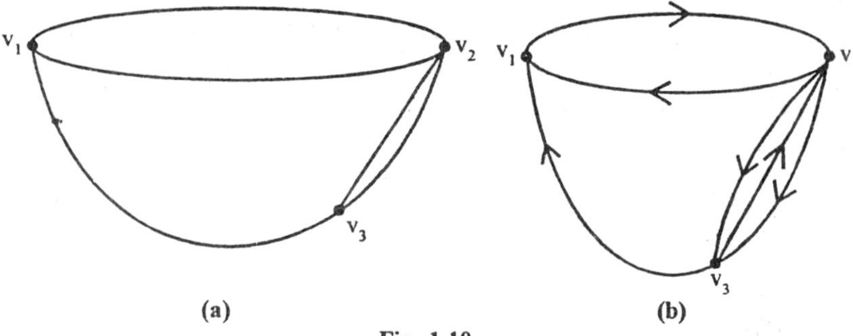

Fig. 1.10

ISOMORPHISM

Two graphs G and G^1 are said to be *Isomorphic* to each other if there is a one–to–one correspondence between their vertices and between their edges such that the incidence relationship is presemed. In other words, if edge e is incident on vertices v_1 and v_2 in G, then the corresponding edge e' in G' must be incident on the vertices v_1 and v_2 that correspond v_1 and v_2 respextively.

SUBGRAPHS

The concept of subgraph is synonym to concept of subset in set theory. A graph g is said to be a *subgraph* of a graph G if all the vertices and all the edges of g are in G, and each edge of g has the same end vertices in g as in G. For example the graph in Fig. 1.11(b) is a sub-graph of Fig. 1.11(a).

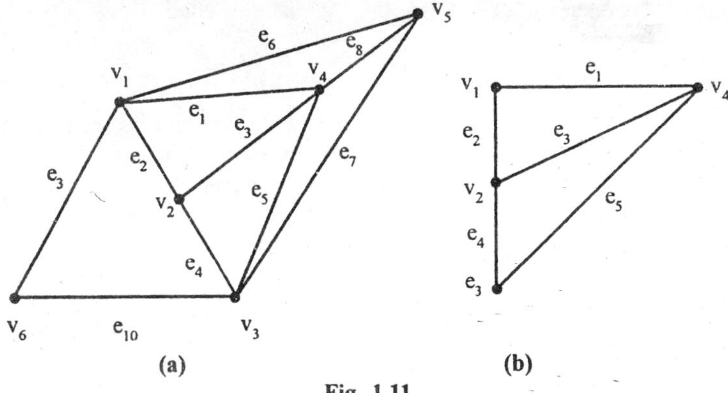

(a) (b)

Fig. 1.11

The symbol, $g \subset G$, is used in stating "g is a subgraph of G".

We can have following observations about the subgraphs.

(i) Every graph is its own subgraph, i.e., $G \subset G$

(ii) A subgraph of a subgraph of G is a subgraph of G.

(iii) A single vertex in a graph G is a subgraph of G.

(iv) A single edge in G, together with its end vertices, is also a subgraph of G.

Edge-Disjoint Subgraphs

Two or more subgraphs g_1 and g_2 of a graph G are said to be *edge disjoint* if g_1 and g_2 do not have any edges in common. For example, the two graphs in Figs. 1.12 (b) and 1.12(c) are the edge-disjoint subgrahs of the graph in Fig. 1.12(a).

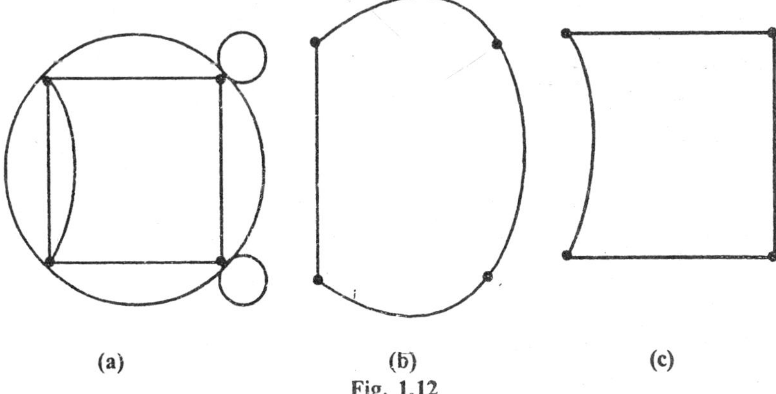

(a) (b) (c)

Fig. 1.12

Although edge-disjoint graphs do not have any edge in common, they may have vertices in common. Subgraphs that do not have vertices in common are said to be *vertex disjoint*.

WALKS, PATHS AND CIRCUITS

A *walk* is a finite alternating sequence of vertices and edges, beginning and ending with vertices, such that each edge is incident with the vertices preceding and following it. No edge appears more than once in a walk, while a vertex may appear more than once. In Fig. 1.13, v_1 a v_2 b v_3 d v_4 e v_2 f v_5 is a walk.

A walk is also referred to as an *edge train* or a *chain*. The set of vertices and edges constituting a given walk in a graph G is clearly a subgraph of G.

Vertices with which a walk begins and ends are called its *terminal vertices*. It is possible for a walk to begin and end at the same vertex. Such a walk is called a *closed walk*. A walk that is not closed in called an *open walk*.

An open walk in which no vertex appears more than once is called a *path* or a *simple path* or an *elementary path*. In Fig. 1.13, v_1 a v_2 b v_3 d v_4 is path whereas v_1 a v_2 b v_3 d v_4 e v_2 f v_5 is not a path. A path does not intersect itself. The number of edges in a path is called the *length of path*.

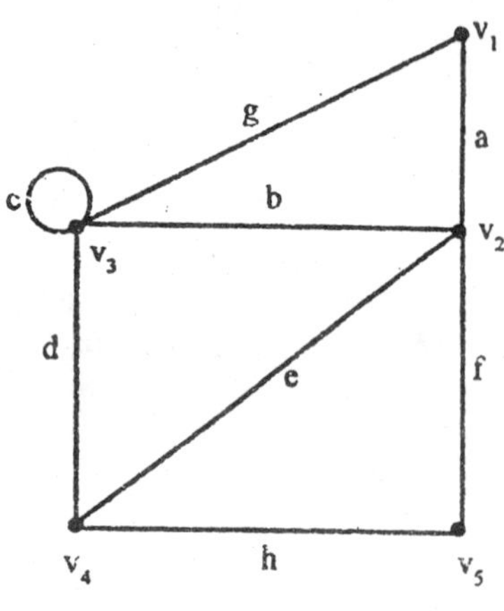

Fig. 1.13

A closed walk in which no vertex, except the initial and the final vertex, appears more than once is called a *circuit*. A circuit is a closed, non-intersecting walk. In Fig. 1.14, we have two circuits.

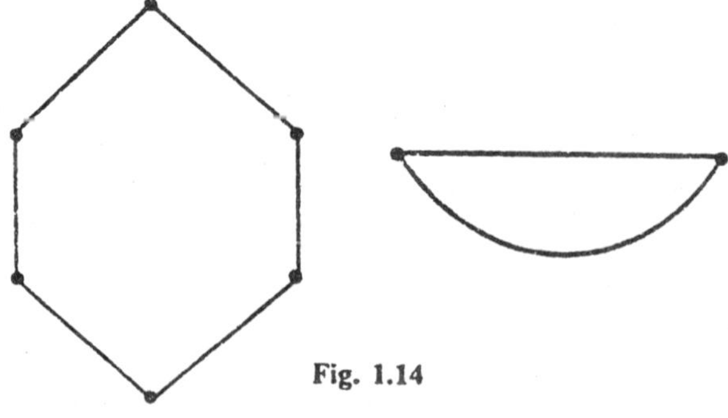

Fig. 1.14

A circuit is also called a *cycle, elementary cycle, circular path,* and *palygon.*

CONNECTED GRAPHS, DISCONNECTED GRAPHS AND COMPONENTS

A graph G is said to be *connected* if there is atleast one path between every pair of vertices in G. Otherwise G is *Disconnected.* For example, the graph in Fig. 1.15 is connected but the graph shown in Fig. 1.15 is disconnected.

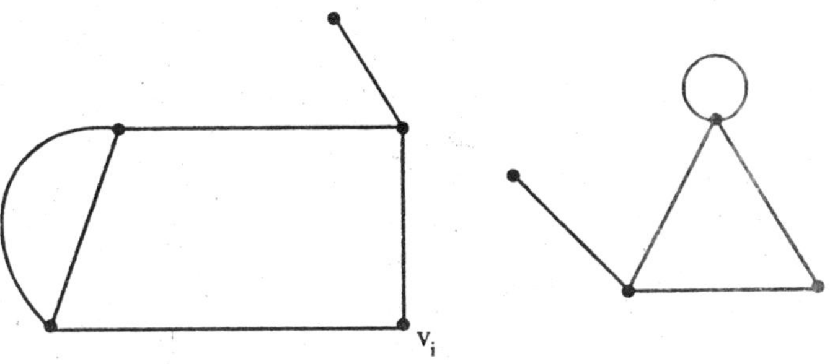

Fig. 1.15

A null graph of more than one vertex is disconnected. Obviously, a disconnected graph consists of two or more connected graphs. Each of these connected subgraphs is called a *component*. For example, the disconnected graph in Fig. 1.15 consists of two components.

Consider a vertex v_i in a disconnected graph G. By definition, not all vertices of G are joined by paths to v_i. Vertex v_i and all the vertices of G that have paths to v_i, together with all the edges incident on them, form a component. Clearly, a component itself is a graph.

Theorem 1:

If a graph (connected or disconnected) has exactly two vertices of odd degree, there must be a path joining these two vertices.

Let G be a graph with all even vertices except vertices v_1 and v_2, which are odd. As we know that the number of vertices of odd degree in a graph is always even, therefore for every compoment of a disconnected graph, no graph can have an odd number of odd vertices. Therefore, in graphy G, v_1 and v_2 must belong to the same component. Hence there must exist a path between them.

Theorem 2:

A simple graph with n vertices and k components can have atmost $(n - k)(n - k + 1)/2$ edges.

Let a graph G has k components. Let the number of vertices in these k components be n_1, n_2, ..., n_k. Then

$$n_1 + n_2 + ... + n_k = n\ n_i \geq 1.$$

The maximum number of edges in the ith component of G, which is a simple connected graph, is 1/2 n_i (n_i – 1). Therefore, the maximum number of edges in G is

$$\frac{1}{2}\sum_{i=1}^{k}(n_i - 1)\, n_i = \frac{1}{2}\left(\sum_{i=1}^{k} n_i\right)^2 - \frac{n}{2} \qquad ...(1)$$

We know the algebraic inequality

$$\sum_{i=1}^{k} n_i^2 \leq n^2 - (k-1)(2n-k) \qquad ...(2)$$

By (2), eqn. (1) becomes

$$\frac{1}{2}\sum_{i=1}^{k}(n_1 - 1)\, n_i \leq \frac{1}{2}[n^2\ (k-1)(2n-k)] - \frac{n}{2}$$

$$= \frac{1}{2}.(n-k)(n-k+1)$$

Theorem 3:

A graph G is disconnected if and only if its vertex set V can be partitioned into two non-empty, disjoint subsets v_1 and v_2 such that there exists no edge in G whose one end vertex is in subset v_1 and the other in subset v_2. Let such a partitioning exists. Let there be two arbitrary vertices a and b of G, such that a ∈ v_1 and b ∈ v_2. No path can exist between vertices a and b; otherwise, there would be atleast one edge whose one end vertex would be in v_1 and the other in v_2. Hence, if such a partition exist, G is not connected.

Conversely, let G be a disconected graph. Let there be a vertex a in G. Let v_1 be the set of all vertices that are joined by paths to a. Since G is disconnected, v_1 does not include all vertices of G. The remaining vertices will form a nonempty set v_2. No vertex in v_1 is joined to any in v_2 by an edge. clearly v_1 and v_2 are disjoint sets. Hence the partition.

EULERGRAPHS

If some closed walk in a graph contains all the edges of the graph, then the walk is called an *Euler line* and the graph an *Euler* graph.

Since the Euler line (which is a walk) contains all the edges of the graph, an Euler graph is always connected except for any isolated vertices. Siince isolated vertices do not contribute any thing to the Euler graph, it is assumed that Euler graphs do not have any isolated vertices.

Theorem:

A given connected graph G is an Euler graph if and only if all vertices of G are of even degree.

Let G be an Euler graph. It therefore contains a closed walk known as Euler line. In tracing this walk we observe that every time the walk meets a vertex v it goes through two new edges incident on v with one it enters v and with the other it exits. This is true not only for all intermediate vertices of the walk but also of the terminal vertex. Thus if G is an Euler graph, the degree of every vertex is even.

For *sufficiency,* assume that all vertices of G are of even degree. Now we construct a walk starting at an arbitrary vertex v and going through the edges of G such that no edge is traced more than once. Since every vertex is of even degree, we can exit from every vertex we enter. The tracing cannot stop at any vertex but v. Since v is also of even degree, we shall reach v when the tracing comes to an end.

If this closed walk h we just traced includes all the edges of G, G is an Euler graph. If not, we remove from G all the edges in h and obtain a subgraph h' of G formed by the remaining edges. Since both G and h have all their vertices of even degree, the degrees of the vertices of h' are also even, h' must touch h at least at one vertex v_1, because G is connected. Starting from v_1, we can again construct a new walk in graph h'. Since all the vertices of h' are of even degree, this walk in h' must terminate at vertex v_1. But this walk in h' can be combined with h to from a new walk, which starts and ends at vertex v and has more edges than h. This process can be repeated until we obtain a closed walk that traverses all the edges of G. Thus G is an Euler graph. Fig. 1.16 represents two Euler graphs.

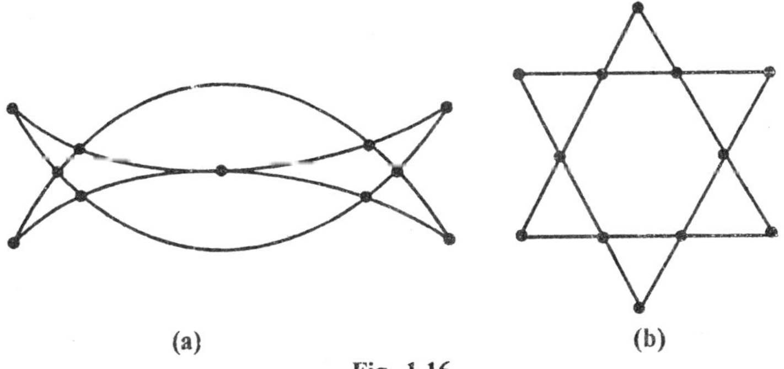

Fig. 1.16

Unicursal Graph

An open walk that includes (or traces or convers) all edges of a graph without retracing any edge is called a *unicursal line* or an *open.*

Euler Line

A connected graph that has a unicursal line will be called a *unicursal graph* Fig. 1.17.

Clearly, by adding an edge between the initial and final vertices of a unicursal line we get an Euler line. Thus a connected graph is unicursal if an only if it has exactly two vertices of odd degree.

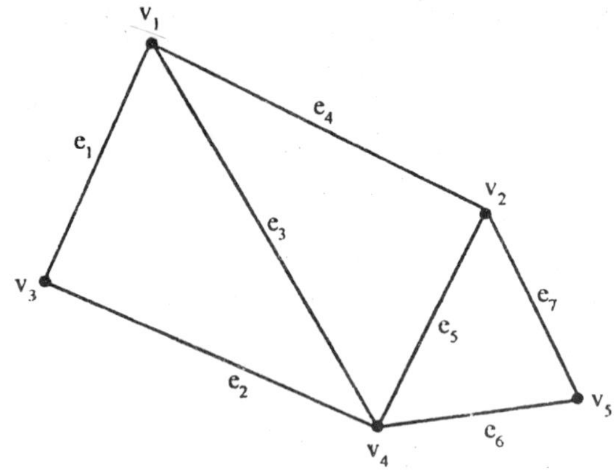

Fig. 1.17

Theorem:

A connected graph G is an Euler graph if and only if it can be decomposed into circuits.

Let graph G can be decomposed into circuits, i.e., G be a union of edge-disjoint circuits. Since the degree of every vertex in a circuit is two, the degree of every vertex in G is even. Hence G is an Euler graph.

Conversely

Let G be an Euler graph. Consider a verrtex v_1. There are atleast edges incident at v_1. Let one of these edges be between v_1 and v_2. Since vertex v_2 is also of even degree, it must have atleast another edge, say between v_2 and v_3. Proceeding in this way, we arrive at a vertex that has previously been traversed, thus forming a circuiit C.

Let us remove C from G. All vertices in the remaining graph must also be of even degree. From the remaining graph remove another circuit in exactly the same way as we removed C from G. Continue this process until no edges are left. Hence the theorem.

HAMILTONIAN CIRCUITS

A circuit in a graph G that contains each vertex in G exactly once, except for the starting and ending vertex that apears twice is known as *Hamiltonian circuit.*

A graph G is called a *Hamiltonian graph* if it contains a Hamilitonian circuit.

A *Hamiltonian path* is a simple path that contains all vertices of G where the end points may be distinct.

Clearly, a Hamiltonian path in a graph G traverses every vertex of G. Since a Hamiltonian path is a subgraph of a Hamiltonian circuit, every graph that has a Hamiltonian circuit also has a Hamiltonian path.

The length of a Hamiltonian path (if it exists) in a connected graph of n vertices is n–1.

Labelled and Weighted Graphs

A graph G is called a *labeled graph* if its edges and/or vertices are assigned data of one kind or another.

A graph G is called a *weighted graph* if each edge e of G is assigned a non-negative number we called the *weight* or *length* of v, weighted graph where the weight of each edge is given in the obvious way.

The *weight* or length of a path in such a weighted graph G is defined to be the sum of the weights of the edges in the graph. The path of minimum weight between any two given vertices is called a *shortest path.* The length of a shortest path between A and B one such path is

$$(A, A_1, A_2, A_5, A_3, A_6, B)$$

Regular Graphs

A graph G is *reqular of degree k* or *k-regular* if every vertex has degree k. In other words, a graph is regular if every vertex has the same degree.

The connected 0-regular graph is the trivial graph with one vertex and no edges. The connected 1-regular graph is the graph wiith two vertices and one edge connecting them.

The connected 2-regular graph with n vertices is the graph which consists of a single n-cycle. [Fig. 1.18].

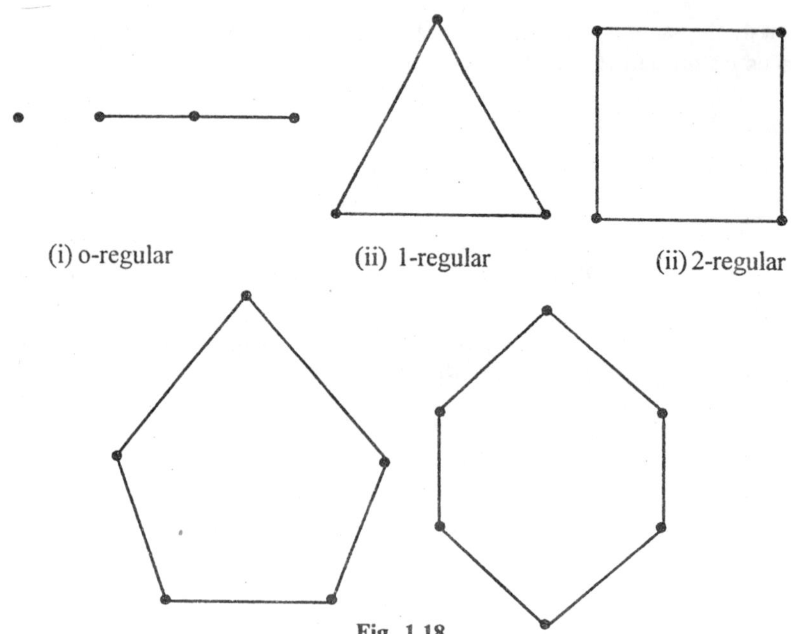

Fig. 1.18

CUT-SETS

In a connected graph G, *cut-set* is a set of edges whose removal from G leaves G disconnected, provided removal of no proper subset of these edges disconnects G.

For example, in Fig. 1.19 the set of edges [e_1, e_3, e_4, e_6] is a cut-set. There are many other cut-sets, such as [e_1, e_2, e_7], [e_1, e_2, e_5, e_6], and [e_4, e_8, e_6]. Edge [e_9] alone is cut-set. The set of edges [e_1, e_3, e_8, e_4], on the other hand, is not a cut-set, because one of its proper subsets [e_1, e_3, e_8] is a cut-set.

A cut-set always 'cuts' a graph into two. Therefore, a cut-set can also be defined as a minimal set of edges in a connected graph whose removal reduces the rank of the graph by one.

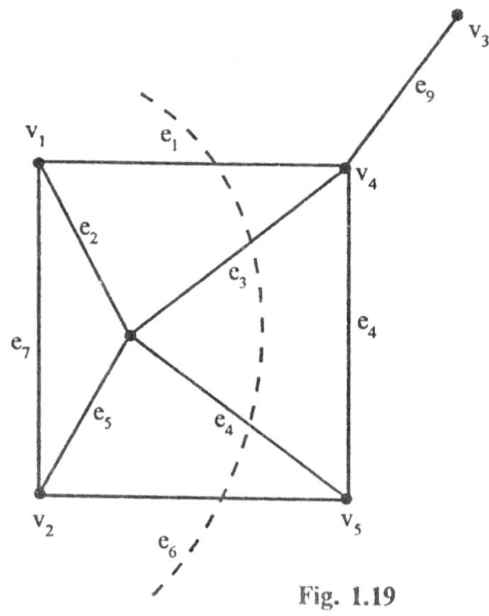

Fig. 1.19

PLANAR GRAPHS

A graph G is said to be *planar* if there exists some geometric representation of G which can be drawn on a plane such that no two of its edges intersect. A graph that cannot be drawn on a plane without a cross over between its edges is called *nonplanar*.

A drawing of a geometric representation of a graph on any surface such that no edges intersect is called *embedding*. Thus a geometric graph G is planar if there exists a graph isomorphic to G that is embedded in a plane. Otherwise, G is nonplanar.

Region

A plane representation of a graph divides the plane into *regions* or *windows*, or *faces* or *meshes* [Fig. 1.20]. In Fig. 1.20, the numbers stand for regions. A region is characterized by the set of edges or the set of vertices forming its *boundary*.

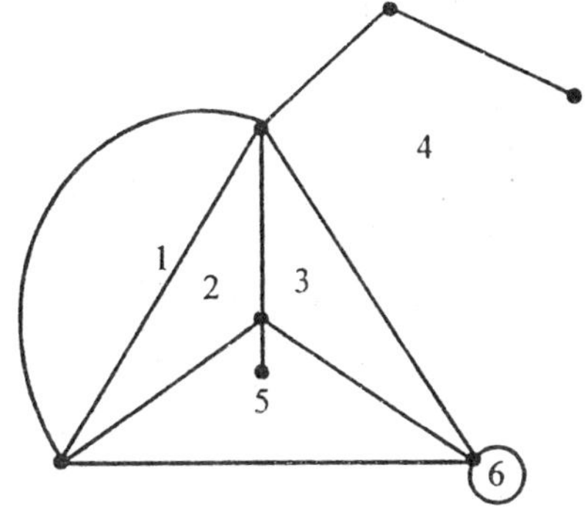

Fig. 1.20

Infinite Region

The portion of the plane lying outside a graph embedded in a plane, such as region infinite in its extent. Such a region is called the *infinite, unbounded, outer* or *exterior* region for that particular plane representation. Any region can be made the infinite region by proper embedding.

Euler's Formula

A connected planar graph with n vertices and e edges has e – n + 2 regions.

Any simple planar graph can have a plane representation such that each edge is a straight line. Any planar graph can be drawn such that each region is a polygon. Let the ploygonal net representing the given graph consist of f regions or faces, and let k_m be the number of m-sided regions. Since each edge is on the boundary of exactly two regions.

$$3 \cdot k_3 + 4 \cdot k_4 + 5 \cdot k_5 + \ldots + m.k_m. = 2 \cdot e \quad \ldots(1)$$

where k_m is the number of polygons, with maximum edges.

$$\text{Also,} \quad k_3 + k_4 + k_5 + \ldots + k_m = f \quad \ldots(2)$$

The sum of all angles subtended at each vertex in the polygonal net

$$= 2\pi\, n \quad \ldots(3)$$

We know that the sum of all interior angles of a m-sided polygon is π (m – 2) and the sum of the exterior angles is π (m + 2). Then sum of all interior angles of f – 1 finite regions plus the sum of the exterior angles of the polygon defining the infinite region is

$$\pi (3-2) \cdot k_3 + \pi (4-2) \cdot k_4 + \ldots + \pi (r-2) \cdot k_2 + 4\pi$$

$$= \pi (2e - 2f) + 4\pi. \qquad \ldots(4)$$

From eqns. (3) and (4), we get

$$2\pi (e - f) + 4\pi = 2\pi n, \quad \text{or} \quad e - f + 2 = n.$$

∴ The number of regions is $f = e - n + 2$.

Corollary *: In any simple, connected planar graph with f regions, n vertices, and e edges (e > 2), the following inequalities must hold:*

$$e \geq 3/2\, f \qquad e \leq 3n - 6$$

Since each region is bounded by atleast three edges and each edge belongs to exactly two regions.

$$2e \geq 3f \quad \text{or} \quad e \geq 3/2\, f. \qquad \ldots(1)$$

Substituting value of f from Euler's formula in (1), we get

$$e \geq 3/2\,(e - n + 2) \quad \text{or} \quad e \leq 3n - 6.$$

CONNECTIVITY AND SEPARABILITY

Edge Connectivity

Each cut-set of a connected graph G consists of a certain number of edges. The number of edges in the smallest cut-set is defiined as the *edge connectivity* of G. In other words, the edge connectivity of a connected graph can be defined as the minimum number of edges whose removal reduces the rank of the graph by one.

Vertex Connectivity

The *vertex connectivity* or simply *connectivity* of a connected graph G is defined as the minimum number of vertices whose removal from G leaves the remaining graph disconnected. The vertex connectivity of the graph.

Seperable Graph

A connected graph is said to be *seperable* if its vertex connectivity is one. All other connected graphs are called *nonseperable*. In other words a connected graph G is said to be seperable if there exists a subgraph g in G such that $\overline{g}$ (the complement of g in G) and g have only one vertex common. In a seperable graph a vertex whose removal disconnects the graph is called a *cut-vertex*, a *cut-node* or an *articulation point*.

HOMEOMORPHIC GRAPHS

Two graphs are said to be *homeomorphic* if one graph can be obtained from the other by the creation of edges in series, i.e., by insertion of vertices of degree two, or by the merger of edges in series. The graphs shown in Fig. 1.21 are homeomorphic to each other.

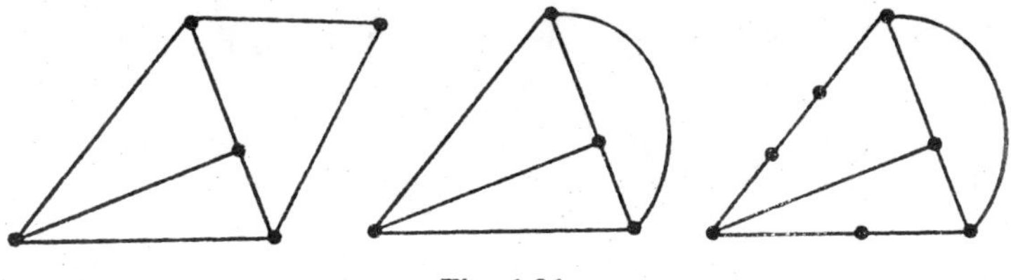

Fig. 1.21

DUAL

Consider the plane representation of a graph in Fig. 1.22(a), with six regions R_1, R_2, R_3, R_4, R_5 and R_6. Let us place six points a_1, a_2, ..., a_6 one in each of the region, as shown in Fig. 1.22(b). Let us join six points according to the following procedure:

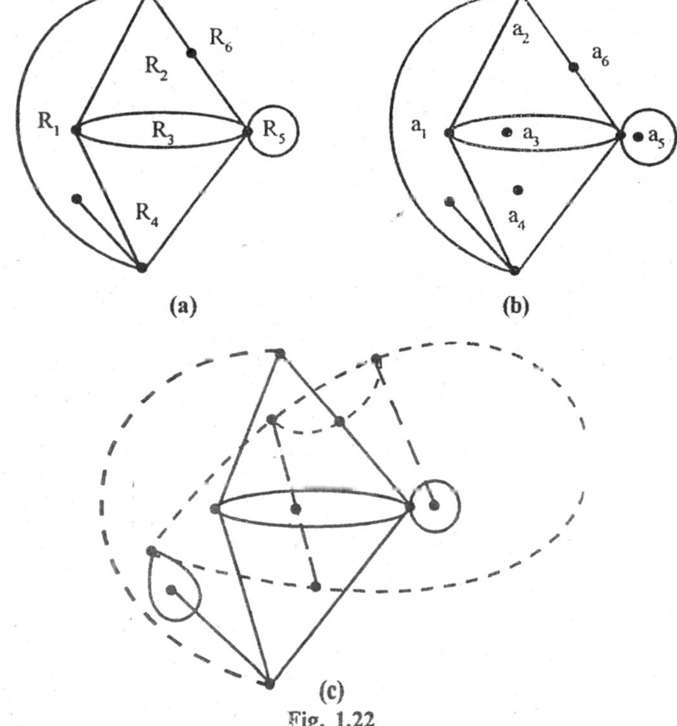

Fig. 1.22

(i) If two regions R_i and R_j are adjacent i.e., have a common edge, draw a line joining the points a_i and a_j that intersects the common edge between R_i and R_j exactly once.

(ii) If there is more than one edge common between R_i and R_j, draw one line between points a_i and a_j for each of the common edge.

(iii) For an edge e lying entirely in one region, say R_k, draw a self-loop at point a_k intersecting e exactly once.

Thus we obtan a new graph G* consisting of six vertices, a_1, a_2, ..., a_6 and edges joining these vertices. Such a graph G* is called a *dual* or a *geometric dual* of G.

It is clear that there is a one-to-one correspondence between the edges of a graph G and its dual G*-one edge of G* intersecting one edge of G. The relationship between a planar graph G and its dual G* can be summed up as follows:

(i) An edge forming a self-loop in G yields a pendant edge in G*.

(ii) A pendant edge in G yields a self-loop in G*.

(iii) Parallel edges in G produce edges in series in G* while the edges that are in series in G produce parallal edges in G*.

(iv) Graph G* is also embedded in the plane and is therefore planar.

Now, if we observe the process of drawing a dual G* from G, it is evident that G is a dual of G*. Therefore, instead of calling G* a dual of G, we usually say that G and G* are dual graphs.

Dual of a Subgraph

Let G be a planar graph and G* be its dual. Let e be an edge in G, and the corresponding edge in G* be e*. Suppose that we delete edge e from G and then try to find the dual of G - e. If edge e was on the boundary of two regions, removal of e would merge these two regions into one.

Thus the dual (G – e)* can be obtained from G* by deleting the corresponding edge e* and then fusing the two end vertices of e* in G* – e*. But, if edge e is not on the boundary, e* forms a self-loop. In this case G* – e* is the same as (G – e)*. Thus dual of any subgraph of G can be obtained by successive application of the above process.

Dual of Honeomorphic Graph

Let G be a planar graph and G* be its dual. Let e be an edge in G whose corresponding edge in G* be e*. Suppose we create an additional vertex in G by introducing a vertex of degree two in edge e. Due to this an edge parallel to e* will add in G*. Similarly, the revesse process of merging two edges in series will eliminate one of the corresponding parallel edges in G*. By this process, if a graph G has a dual G*, the dual of any graph homeomorphic to G can be obtained from G*.

OPERATIONS OF GRAPHS

Now, we will discuss some important operations on graphs:

Union

Given two graphs G_1 and G_2. Their union will be a graph such that

$$V(G_1 \cup G_2) = V(G_1) \cup V(G_2)$$

and

$$E(G_1 \cup G_2) = E(G_1) \cup E(G_2)$$

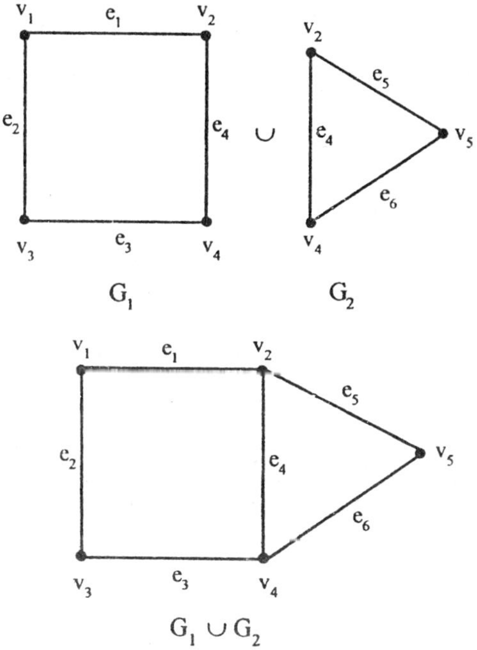

v_1 e_1 v_2 e_2 e_4 v_3 e_3 v_4 $\cup$ v_2 e_5 e_4 v_5 e_6 v_4

G_1 G_2

v_1 e_1 v_2 e_5 e_2 e_4 v_5 e_6 v_3 e_3 v_4

$G_1 \cup G_2$

Fig. 1.23

Intersection

Given two graphs G_1 and G_2 with at least one vertex in common then their intersection will be a graph such that

$$V(G_1 \cap G_2) = V(G_1) \cap V(G_2)$$

and

$$E(G_1 \cap G_2) = E(G_1) \cap V(G_2)$$

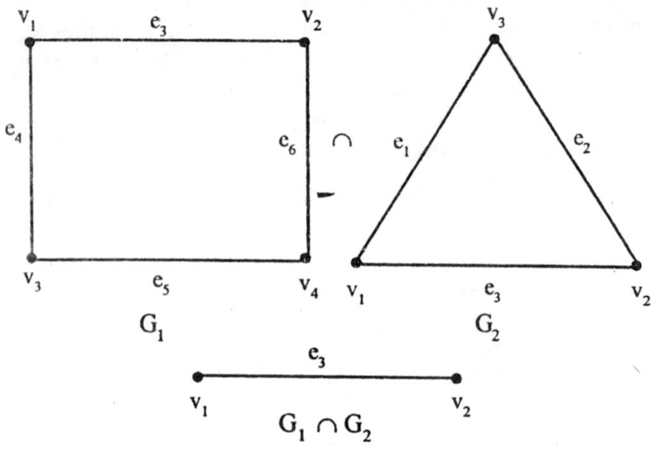

Fig. 1.24

Complement

The complement G' of G is defined as a simple graph with the same vertex set as G and where two vertices u and v are adjacent only when they are not adjacent in G.

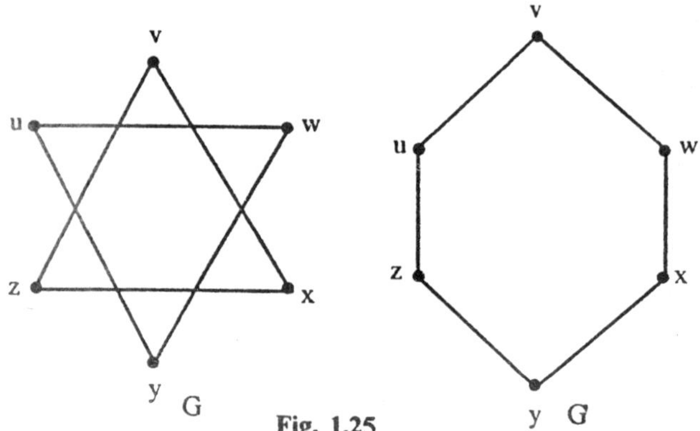

Fig. 1.25

Remark :

(i) Operations of union, intersection and ring sum on two graphs are commutative, i.e.,

$$G_1 \cup G_2 = G_2 \cup G_1;$$
$$G_1 \cap G_2 = G_2 \cap G_1;$$
and $$G_1 \oplus G_2 = G_2 \oplus G_1;$$

(ii) If G_1 and G_2 are edge disjoint, then $G_1 \cap G_2$ is a null graph, and $G_1 \oplus G_2 = G_1 \cup G_2$. If G_1 and G_2 are vertex disjoint, then $G_1 \cap G_2$ is empty.

(iii) For any graph G,

$$G \cup G = G \cap G = G$$

and $$G \oplus G = \text{a null graph.}$$

(iv) If g is a subgraph of G, then $G \oplus g$ is that subgraph of G which remains after all the edges in g have been removed from G. Therefore, $G \oplus g$ is written as $G - g$, whenever $g \subseteq G$. Due to this $G \oplus g = G - g$ is often called the complement of g is G.

Decomposition

A graph is said to have been *decomposed* into two subgraphs g_1 and g_2 if

$$g_1 \cup g_2 = G \quad \text{and} \quad g_1 \cap g_2 = \text{a null graph}$$

Every edge of G occurs either in g_1 or in g_2, but not in both. Some of the vertices, however, may occur in both g_1 and g_2. In decomposition, isotated vertices are disregarded. A graph containing m edges can be decomposed in $2^{m-1} - 1$ different ways into pairs of subgraphs g_1, g_2.

Deletion

Let v_i is a vertex in graph G. $G - v_i$ denotes a subgraph of G obtained by deleting v_i from G. Deletion of a vertex always implies the deletion of all edges incident on that vertex. If e_j is an edge in G, then $G - e_j$ is a subgraph of G obtained by deleting e_j from G. Deletion of an edge does not imply deletion of its end vertices. Hence $G - e_j = G \oplus e_j$.

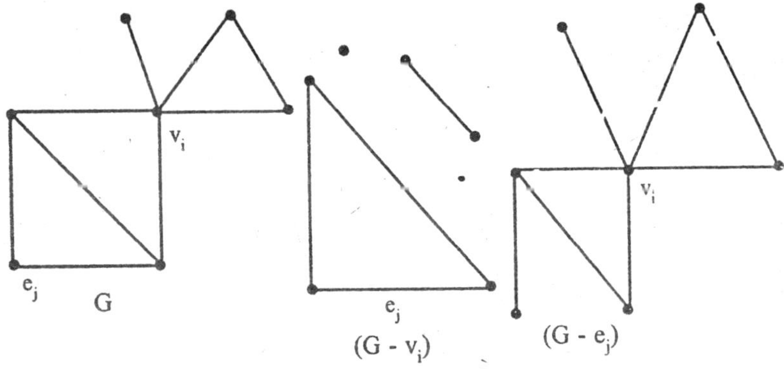

Fig. 1.26

Fusion

A pair of vertices v_1, v_2 in a graph are said to *fused (merged* or *identified)* if the two vertices are replaced by a single new vertex such that every edge that was incident on either v_1 or v_2 or on both is incident on the new vertex. Thus fusion of two vertices does not alter the number of edges, but it reduces the number of vertices by one [Fig. 1.27].

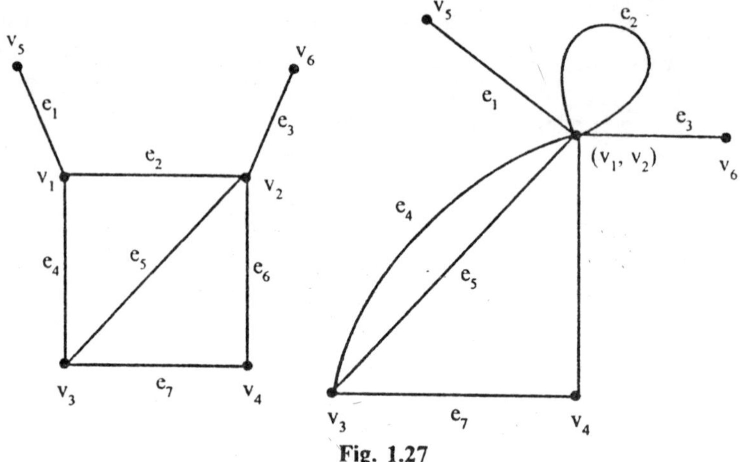

Fig. 1.27

Product of Graphs

To define the product $G_1 \times G_2$ of two graphs consider any two points $u = (u_1, u_2)$ and $v = (v_1, v_2)$ in $V = V_1 \times V_2$. Then u and v are adjacent in $G_1 \times G_2$ whenever $[u_1 = v_1$ and u_2 adj $v_2]$ or $[u_2 = v_2$ and u_1 adj $v_1]$.

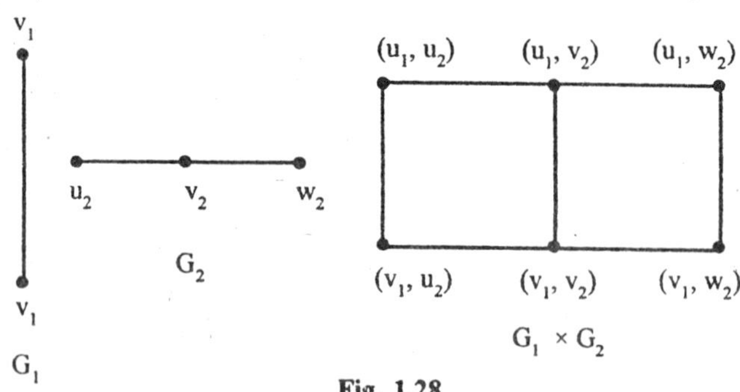

Fig. 1.28

Composition

The composition $G = G_1 [G_2]$ also has $V = V_1 \times V_2$ as its point set, and $u = (u_1, u_2)$ is adjacent with $v = (v_1, v_2)$ whenever $[u_1$ adj $v_1]$ or $[u_1 = v_1$

and u_2 adj v_2]. For the graphs of Fig. 1.29, both composition G_1 [G_2] and G_2 [G_1] are shown in Fig. 1.29.

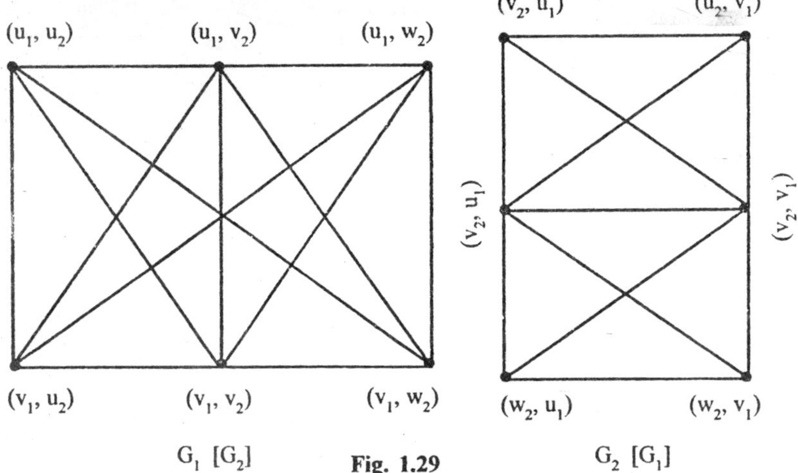

Fig. 1.29

MATRIX REPRESENTATION OF GRAPHS

Although a diagrammatic representation of a graph is very convenient for a visual study, other representations, e.g., matrix representation, are better for computer processing.

Many known results of matrix algebra can be applied to study the structural properties of graphs from an algebraic point of view. In many applications of graph theory, such as in electrical network analysis and operations research matrics proved to be the most powerful tools in expressing the problem.

Incidence Matrix

Suppose G is a graph with n vertices, e edges and no self-loops. Define an n × e matrix **A** = [a_{ij}], whose n rows corresponed to the n vertices and the e columns corrsponed to the e edges, as follows:

a_{ij} = 1, if jth edge e_j is incident on ith vertex v_i, and

= 0 otherwise.

Such a matrix **A** is called the *vertex edge incidence matrix* or simply *incidence matrix.*

Sometimes matrix **A** for a graph G is written as **A**(G). Fig. 1.30 represents a graph G and its incidence matrix.

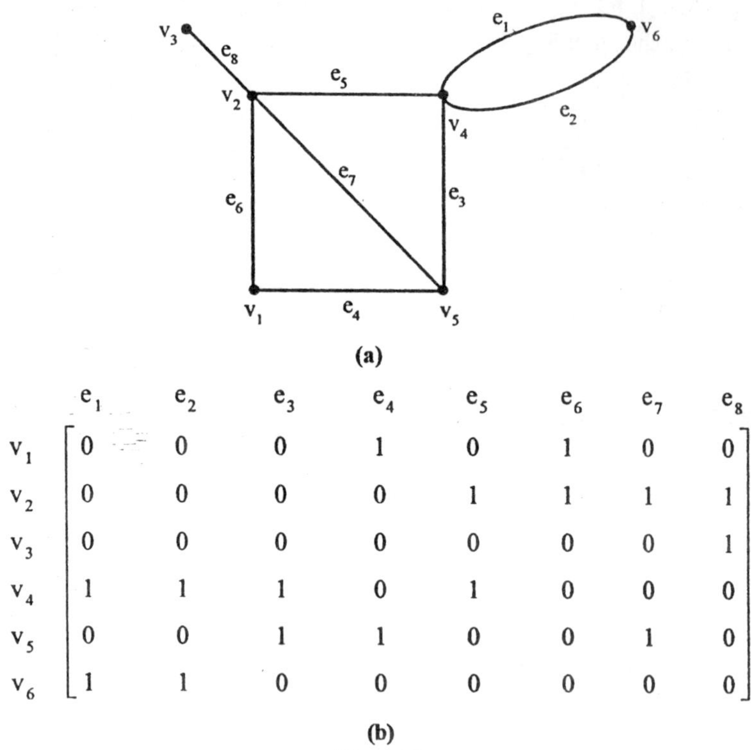

(a)

$$
\begin{array}{c} \\ v_1 \\ v_2 \\ v_3 \\ v_4 \\ v_5 \\ v_6 \end{array}
\begin{array}{c}
\begin{array}{cccccccc} e_1 & e_2 & e_3 & e_4 & e_5 & e_6 & e_7 & e_8 \end{array} \\
\begin{bmatrix}
0 & 0 & 0 & 1 & 0 & 1 & 0 & 0 \\
0 & 0 & 0 & 0 & 1 & 1 & 1 & 1 \\
0 & 0 & 0 & 0 & 0 & 0 & 0 & 1 \\
1 & 1 & 1 & 0 & 1 & 0 & 0 & 0 \\
0 & 0 & 1 & 1 & 0 & 0 & 1 & 0 \\
1 & 1 & 0 & 0 & 0 & 0 & 0 & 0
\end{bmatrix}
\end{array}
$$

(b)

Fig. 1.30

The incidence matrix contains only two elements, 0 and 1. Hence, such a matrix is called a *binary matrix* or a (0, 1) - *matrix*.

By close study of the incidence matrix **A**, we can have following observations:

(i) Since every edge is incident on exactly two vertices, each column of **A** has exactly two.

(ii) In each row number of 1's is equal to the degree of the corresponding vertex.

(iii) Therefore, an isolated vertex is represented by a row with all 0's.

(iv) Parallel edges in a graph produce idential columns in its incidence matrix. For example, columns 1 and 2 in Fig. 1.31(b).

(v) If a graph G is disconnected and consists of two components g and g', the incidence matrix **A** (G) of graph G can be written in a diagraal from as where A(g) and A(g') are the incidence matrices of components g and g'. This is also true for a disconnected graph with any number of components.

$$A(G) = \left[\begin{array}{c|c} A(g) & O \\ \hline O & A(g') \end{array}\right]$$

Submatrices of A(G)

Let g be a subgraph of a graph G, and let A(g) and A(G) be the incidence matrices of g and G respectively. Clearly, A(g) is a submatrix of A(G).

Infact, there is a one-to-one correspondence between each n by k submatrix of A(G) and a subgraph of G with k edges, k being any positive integer less than e and n being the number of vertices in G.

Circuit Matrix

Let the number of different circuits in a graph G be q and the number of edges in G be e. Then a *circuit matrix* **B** = $[b_{ij}]$ of G is a q × e, (0, 1) – matrix defined as follows:

b_{ij} = 1, if ith circuit includes jth edge, and

= 0, otherwise.

The four different circuits, $[e_1, e_2]$, $[e_3, e_5, e_7]$, $[e_4, e_6, e_7]$, and $[e_3, e_4, e_6, e_5]$. Therefore its circuit matrix is a 4 × 8, (0, 1) – matrix as given below:

$$B(G) = \begin{array}{c} \\ 1 \\ 2 \\ 3 \\ 4 \end{array}\begin{array}{c} \begin{array}{cccccccc} e_1 & e_2 & e_3 & e_4 & e_5 & e_6 & e_7 & e_8 \end{array} \\ \left[\begin{array}{cccccccc} 1 & 1 & 0 & 0 & 0 & 0 & 0 & 0 \\ 0 & 0 & 1 & 0 & 1 & 0 & 1 & 0 \\ 0 & 0 & 0 & 1 & 0 & 1 & 1 & 0 \\ 0 & 0 & 1 & 1 & 1 & 1 & 0 & 0 \end{array}\right] \end{array} \quad ...(1)$$

The following observations can be made about a circuit matrix B(G) of a graph G:

(i) A column of all zeros corresponds to a non-circuit edge, i.e., an edge that does not belong to any circuit.

(ii) Each row of **B**(G) is a circuit vector.

(iii) The row corresponding to a self-loop will have a single 1. (In this eway the circuit matrix differs from the incidence matrix).

(iv) The number of 1's in a row is equal to the number of edges in the corresponding circuit.

(v) If graph G is disconnected or seperable and consists of two components g and g', the circuit matrix **B**(G) can be written in a block-diagonal from as

$$\mathbf{B}(G) = \left[\begin{array}{c|c} \mathbf{B}(g) & O \\ \hline O & \mathbf{B}(g') \end{array}\right]$$

Here **B**(g) and **B**(g') are the circuit matrices of g and g'.

Cut-Set Matrix

A cut-set matrix **C** = $[c_{ij}]$ is defined as a matrix in which the rows correspond to the cut-sets and the columns to the edges of the graph, as follows:

c_{ij} = 1, if ith cut-set contains jth edge, and = 0, otherwise

A graph and its cut-set matrix are shown in Fig. 1.31.

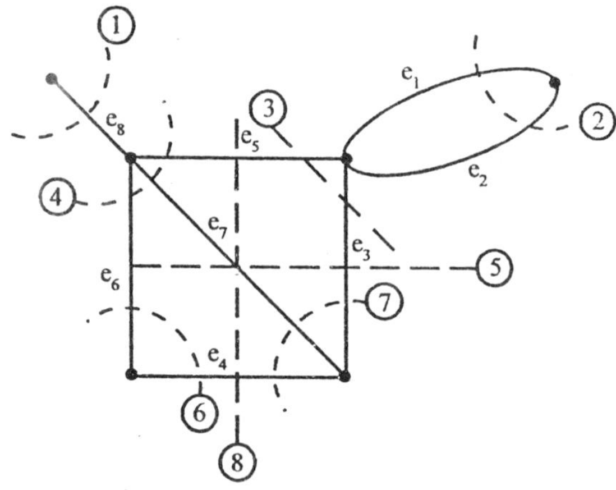

(a)

	e_1	e_2	e_3	e_4	e_5	e_6	e_7	e_8
1	0	0	0	0	0	0	0	1
2	1	1	0	0	0	0	0	0
3	0	0	1	0	1	0	0	0
4	0	0	0	0	1	1	1	0
5	0	0	1	0	0	1	1	0
6	0	0	0	1	0	1	0	0
7	0	0	1	1	0	0	1	0
8	0	0	0	1	1	0	1	0

(b)

Fig. 1.31

We may have following observations about a cut-set matrix **C**(G) of a graph G.

(i) Each row in **C**(G) is a cut-set vector.

(ii) A column with all 0's corresponds to an edge forming a self-loop.

(iii) Parallel edges produce identical columns in the cut-set matrix.

Path Matrix

A path is defined for a specific pair of vertices in a graph, say (x, y), and is written as **P**(x, y). The rows in **P**(x, y) correspond to different paths between vertices x and y, and the columns correspond to the edges in G. That is, the path matrix for (x, y) vertices is **P**(x, y) = $[p_{ij}]$, where

p_{ij} = 1, if jth edge lies in ith path, and = 0, otherwise

Some of the observations we can make about a path matrix **P**(x, y) of a graph G are

(i) A column of all 0's corresponds to an edge that does not lie in any path between x and y.

(ii) A column of all 1's corresponds to an edge that lies in every path between x and y.

(iii) There is no row with all 0's.

(iv) The ring sum of any two rows in **P**(x, y) corresponds to a circuit or an edge-disjoint union of circuits.

Adjacency Matrix

The *adjacency matrix* or *connection matrix* of a graph G with n vertices and no parallel edges is an n × n symmetric binary matrix **X** = $[x_{ij}]$ defined over the ring of integers such that

x_{ij} = 1, if there is an edge between ith and jth vertices, and = 0, if there is no edge between them.

A simple graph and its adjacency matrix are shown in Fig. 1.32.

We can make following observations about the adjacency matrix **X** of a graph G are:

(i) The enteries along the principal diagonal of **X** are all 0's if and only if the graph has no self-loops. A self-loop at the kth vertex corresponds to x_{kk} = 1.

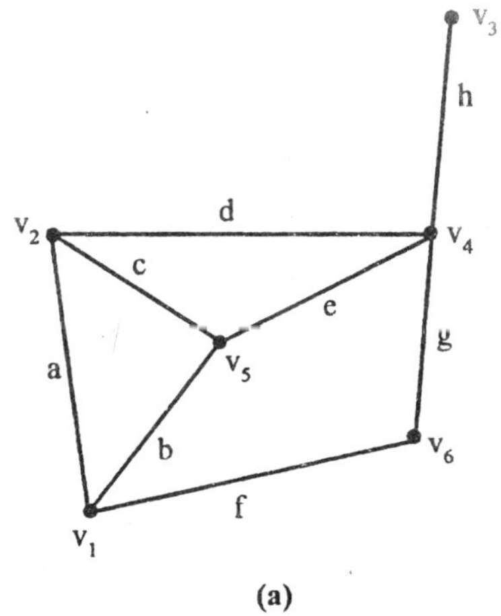

(a)

(ii) There is no provision for parallel edges in adjacency matrix. So why the adjacency matrix **X** is defined for the graphs without parallel edges.

(iii) If the graph has no self-loops, the degree of a vertex equals the number of 1's in the corresponding row or column of **X**.

$$\begin{array}{c} \\ v_1 \\ v_2 \\ v_3 \\ v_4 \\ v_5 \\ v_6 \end{array} \begin{array}{c} \begin{array}{cccccc} v_1 & v_2 & v_3 & v_4 & v_5 & v_6 \end{array} \\ \begin{bmatrix} 0 & 1 & 0 & 0 & 1 & 1 \\ 1 & 0 & 0 & 1 & 1 & 0 \\ 0 & 0 & 0 & 1 & 0 & 0 \\ 0 & 1 & 1 & 0 & 1 & 1 \\ 1 & 1 & 0 & 1 & 0 & 0 \\ 1 & 0 & 0 & 1 & 0 & 0 \end{bmatrix} \end{array}$$

c
Green

(b)
Fig. 1.32

(iv) Two graphs G_1 and G_2 with no parallel edges are isomorphic if and only if their adjacency matrices $\mathbf{X}(G_1)$ and $\mathbf{X}(G_2)$ are related as follows:

$$\mathbf{X}(G_2) = \mathbf{R}^{-1} \,.\, \mathbf{X}(G_1) \,.\, \mathbf{R}_1$$

where **R** is a permutation matrix.

(v) A graph G is disconnected and is in two components g and g' if and only if its adjacency matrix **X**(G) can be partitioned as

$$\mathbf{X}(G) = \left[\begin{array}{c|c} \mathbf{X}(g) & O \\ \hline O & \mathbf{X}(g') \end{array}\right]$$

where **X**(g) is the adjacency matrix of the component g and **X**(g') is that of component g'.

This partitioning implies that there exists no edge joining any vertex in subgraph g to any vertex in subgraph g'.

COLOURING OF GRAPHS

Painting all the vertices of a graph with colours such that no two adjacent vertices have the same colour is called the *colouring* or the *proper colouring* of a graph. A graph in which every vertex has been assigned a colour according to a proper colouring is called a *properly coloured graph.* A given graph can be properly coloured in many different ways. Fig. 1.33 shows three different proper colourings of a graph.

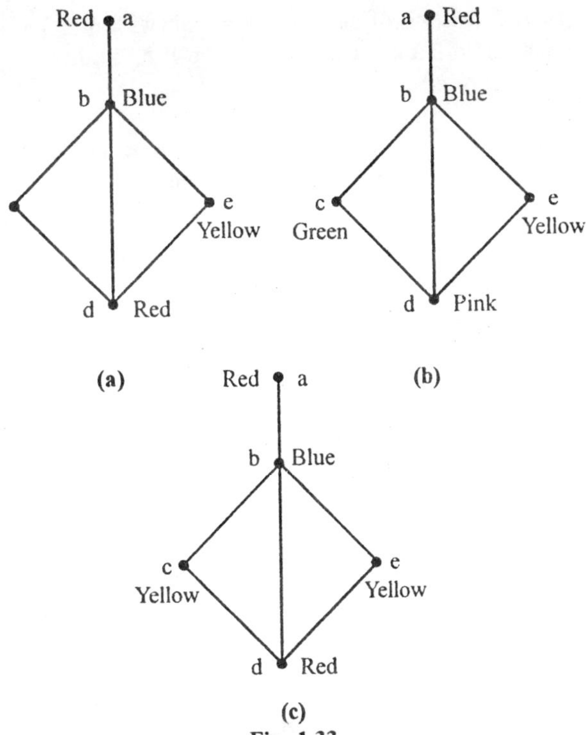

(c)
Fig. 1.33

Chromatic Number

The proper colouring of a graph is one that requires the minimum number of colours. A graph G that requires m different colours for its proper colouring, and no less, is called a *m-chromatic* graph, and the number m is called the *chromatic number* of G. The graphs 3-chromatic.

In colouring graphs, the colouring of disconnected graphs has no meaning. Therefore, it is usual to investigate colouring of connected graphs only. All parallel edges between two vertices can be replaced by a single edge without affecting adjacency of vertices. Self-loops must be disregarded. Thus for colouring of graphs we consider only simple connected graphs.

On colouring of graphs we can list following observations:

(i) A graph consisting of only isolated vertices is 1-chromatic.

(ii) A graph with one or more edges (self-loops disregarded) is at least *2-chromatic* or *bichromatic.*

(iii) A complete graphs of n vertices is n-chromatic, as all its vertices are adjacent.

(iv) A graph consisting of simply one circuit with $n \geq 3$ vertices is 2-chromatic if n is even and 3-chromatic if n is odd.

Matchings

A *matching* or *assignment* in a graph is a subset of edges in which no two edges are adjacent. A single edge in a graph is obviously a matching.

A matching to which no edge in the graph can be added is called *maximal matching*. For example, in a complete graph of three vertices, any single edge is a maximal matching. The edges shown by heavy lines in Fig. 1.34 are two maximal matchings.

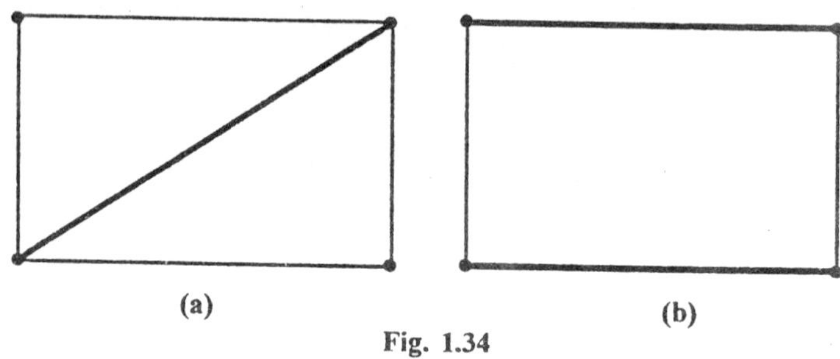

(a) (b)

Fig. 1.34

A graph may have many different maximal matchings, and of different sizes. Among these, the maximal matchings with the largest number of edges are called the *largest maximal matchings.* The number of edges in a largest maximal matching is called the *matching number* of the graph.

Coverings

In a graph G, a set g of edges is said to *cover* G if every vertex in G is incident on at least one edge in g. A set of edges that covers a graph G is said to be an *edge covering,* a *covering subgraph* or *a covering* of G. A graph G is trivially its own covering.

SOME FACTS ABOUT DIRECTED GRAPHS

We have introduced directed graphs or digraphs in chapter 3 and studied then with respect to different relations. Here, we will describe some more facts about the digraphs.

Directed Paths and Connectedness

In walks, paths and circuits in a directed graph, considerations of orientation are also added. For example, in Fig. 1.35, the sequence of vertices and edges $v_5\ e_8\ v_3\ e_6\ v_4\ e_3\ v_1$ is a path directed from v_5 to v_1, where as $v_5\ e_7\ v_4\ e_6\ v_3\ e_1\ v_1$ has no consistent direction from v_5 to v_1.

First one of these paths is known as *directed path* from v_5 to v_1, and the second one is called a *semipath.* The word path in a digraph means either a directed path or a semipath.

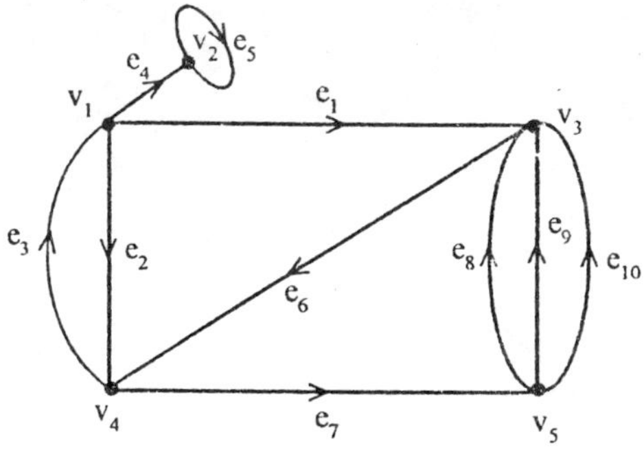

Fig. 1.35

A *directed walk* from a vertex v_i to v_j is an alternating sequence of vertices and edges, beginning with v_i and ending with v_j, such that each edge is oriented from the vertex preceding it to the vertex following it.

A *semi walk* in a directed path is walk in the corresponding unidirected graph, but is not a *directed walk.* A walk in a digraph can mean either a directed walk or a semiwalk.

Similarly, *circuit, semicircuit* and *directed circuit* can be defined.

Connected Digraphs

A digraph G is said to be *strongly connected* if there is at least one directed path from every vertex to every other vertex. A digraph G is said to be *wealkly connected* if its corresponding unidirected graph is connected but G is not strongly connected.

Since there are two types of connectedness in a digraph, we can define two types of components also. Each maximal connected (wealky or strongly) subgraph of a digraph G will be called a *component* of G.

Within each component of G the maximal strongly connected subgraphs will be called the *fragments* of G.

For example, the digraph in Fig. 1.36 consists of two components. The component g contains three fragments $\{e_1, e_2\}$, $\{e_5, e_6, e_7, e_8\}$ and $\{e_{10}\}$. Edges e_3, e_4 and e_9 do not appear in any fragment of g.

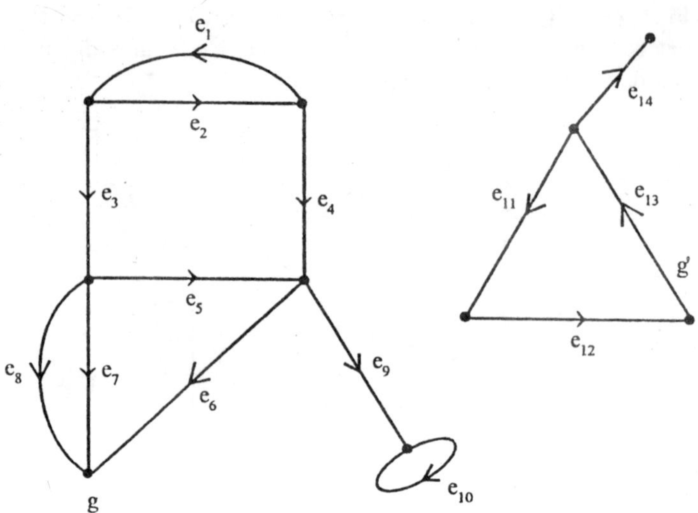

Fig. 1.36

Condensation

The *condensation* G_c of a diagraph G is a digraph in which each strongly connected fragment is replaced by a vertex, and all directed edges from one strongly connected component to another are replaced by a single directed edge. The condensation of the digraph G in Fig. 1.36 is shown in Fig. 1.37. Thus

(i) The condensation of a strongly connected digraph is simply a vertex.

(ii) The condensation of a digraph has no directed circuit.

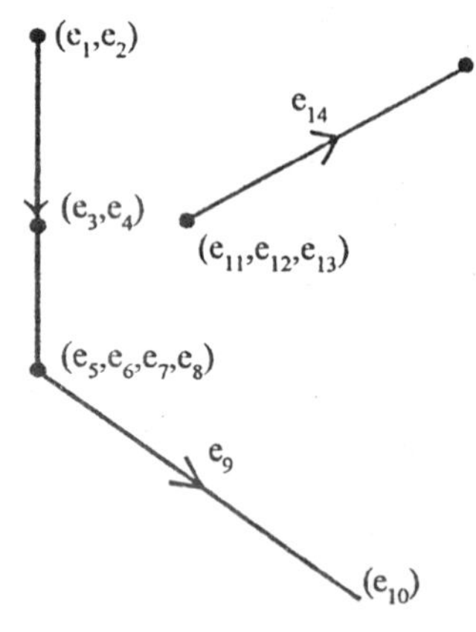

Fig. 1.37

Accessibility

In a diagraph a vertex v_j is said to be *accessible* or *reachable* from vertex v_i if there is a directed path from v_i to v_j. A digraph G is strongly connected if and only if every vertex in G is acessible from every other vertex.

Euler Digraphs

In a digraph G, a closed directed walk which traverses every edge of G exactly once is called a *directed Euler line*. A digraph containing a directed Euler line is called an *Euler digraph.* The graph in Fig. 1.38 is an Euler digraph, in which the walk e_1 e_2 e_3 e_4 e_5 e_6 is an Euler line.

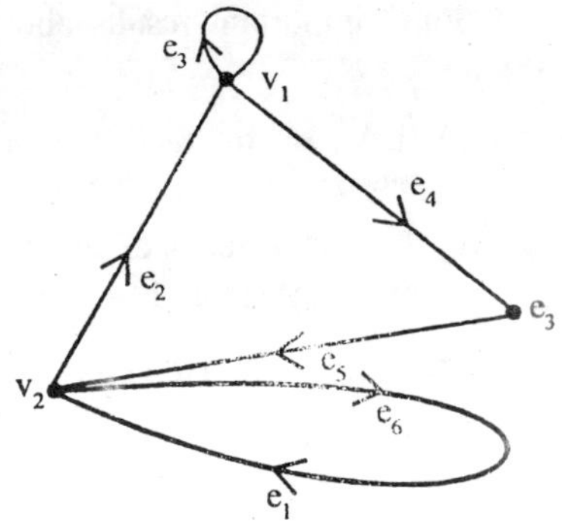

Fig. 1.38

Incidence Matrix of Digraph

The *incidence matrix* of a digraph with n vertices, e edges, and no self-loops is an n × n matrix $A = [a_{ij}]$, whose rows correspond to vertices and columns correspond to edges, such that

a_{ij} = 1, if jth edge is incident out of ith vertex.

= –1, if jth edge is incident into ith vertex.

= 0, if jth edge is not incident on ith vertex.

A digraph and its incidence matrix are shown in Fig. 1.39. Since the sum of each column is zero, the rank of the incidence matrix of a digraph of n vertices is less than n.

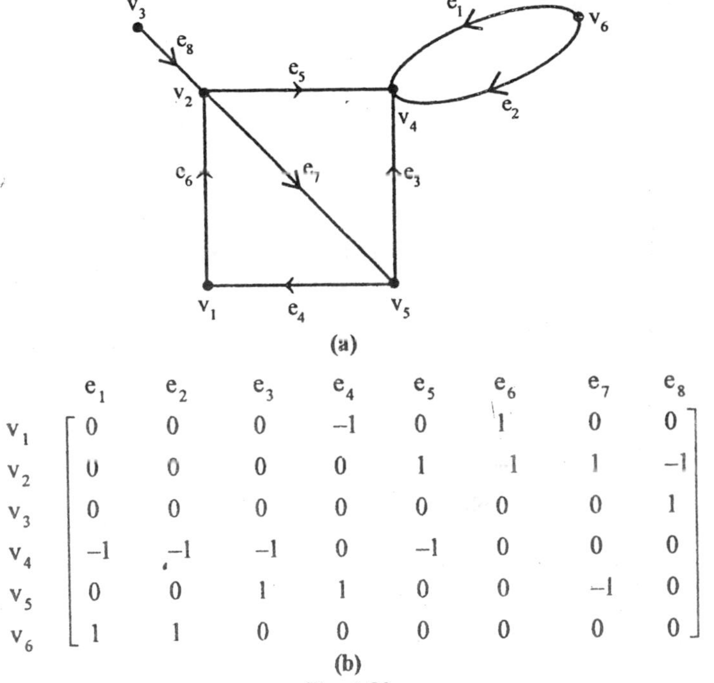

(a)

	e_1	e_2	e_3	e_4	e_5	e_6	e_7	e_8
v_1	0	0	0	–1	0	1	0	0
v_2	0	0	0	0	1	1	1	–1
v_3	0	0	0	0	0	0	0	1
v_4	–1	–1	–1	0	–1	0	0	0
v_5	0	0	1	1	0	0	–1	0
v_6	1	1	0	0	0	0	0	0

(b)

Fig. 1.39

Some important results about incidence matrix of a diagraph are as follows:

(i) If A(g) is the incidence matrix of a connected digraph of n vertices, the rank of A(G) = n – 1.

(ii) The determinant of every square submatrix of A, the incidence matrix of a digraph is 1, –1 or 0.

Circuit Matrix of a Digraph

Let G be a digraph with e edges and q circuits. An arbitrary orientation is assigned to each of the q circuits. Then a circuit matrix $\mathbf{B} = [b_{ij}]$ of the digraph G is a q × e matrix defined as:

b_{ij} = 1, if ith circuit includes jth edge, and the orientations of the edge and circuit coincide.,

= –1, if ith circuit includes jth edge, but the orientations of the two are opposite.

= 0, if ith circuit does not include the jth edge.

For example, a circuit matrix of the digraph in Fig. 10.54 is

$$\begin{bmatrix} e_1 & e_2 & e_3 & e_4 & e_5 & e_6 & e_7 & e_8 \\ 0 & 0 & 0 & 1 & 0 & 1 & 1 & 0 \\ 0 & 0 & 1 & 0 & -1 & 0 & 1 & 0 \\ 0 & 0 & 1 & -1 & -1 & -1 & 0 & 0 \\ -1 & 1 & 0 & 0 & 0 & 0 & 0 & 0 \end{bmatrix}$$

Some important results about incidence matrix A(G) and cicuit matrix B(G) of a digraph G are:

(i) $A \cdot B^T = B \cdot A^T = 0$

where subscript T denotes the transposed matrix.

(ii) Rank of of **B** + rank of **A** = e $\Rightarrow$ rank of **B** = e – n + 1.

Adjacency Matrix of a Digraph

Let G be a digraph with n vertices, containing no parallel edges. Then the *adjacency matrix* $\mathbf{X} = [x_{ij}]$ of the digraph G is an n × n (0, 1) matrix whose elements are as follows:

x_{ij} = 1, if there is an edge directed from ith vertex to jth vertex.

= 0, otherwise.

A digraph and its adjacency matrix are shown in Fig. 1.40.

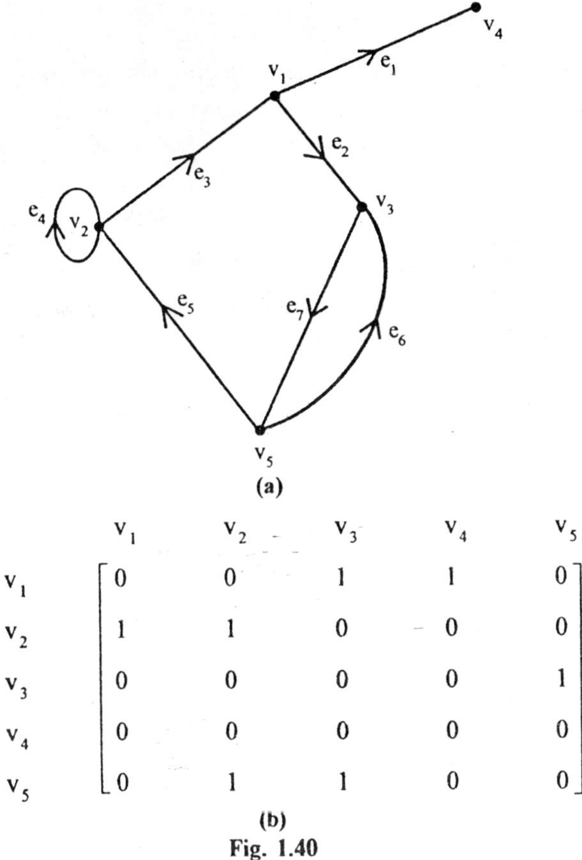

$$
\begin{array}{c c}
 & \begin{array}{ccccc} v_1 & v_2 & v_3 & v_4 & v_5 \end{array} \\
\begin{array}{c} v_1 \\ v_2 \\ v_3 \\ v_4 \\ v_5 \end{array} &
\begin{bmatrix}
0 & 0 & 1 & 1 & 0 \\
1 & 1 & 0 & 0 & 0 \\
0 & 0 & 0 & 0 & 1 \\
0 & 0 & 0 & 0 & 0 \\
0 & 1 & 1 & 0 & 0
\end{bmatrix}
\end{array}
$$

(b)

Fig. 1.40

Adjacency matrix **X** of a digraph G has following properties:

(i) **X** is a symmetric matrix if and only if G is a symmetric digraph.

(ii) Every non-zero element on the main diagonal of **X** represents a self-loop at the corresponding vertex.

(iii) In **X** there is no way of showing parallel edges. Therefore, the adjacency matrix is defined only for a digraph without parallel edges.

(iv) The sum of each row of **X** equals the out-degree of the corresponding vertex, and the sum of the corresponding vertex, and the sum of each column of **X** equals the in-degree of the corresponding vertex. The number of non-zero entries in **X** equals the number of edges in G.

(v) If **X** is the adjacency matrix of a digraph G, then the adjacency matrix of a digraph G^R obtained by reversing the direction of every edge in G is the transposed matrix X^T.

PLANARITY AND COLORINGS

The topics in this section are related to how graphs are drawn.

Planarity—Can a given graph be drawn in an plane so that no edges intersect? Certainly, it is natural to avoid intersections, but up to now we haven't gone out of our way to do so.

Colorings—Suppose that each vertex in an undirected graph is to be colored so that no two vertices that are connected by an edge have the same color. How many colors are needed? This questions is motivated by the problem of drawing a map so that no two bordering countries are colored the same. A similar question can be asked for coloring edges.

Definition: Planar Graph. *A graph is planar if it can be drawn in a plane (e.g., a sheet of paper) so that no edges cross.*

Notes:

(a) *In discussing planarith, we need only consider simple undirected graphs with no self-loops.* All other graphs can be treated as such since all of the edges that relate any two vertices can be considered as one "package" that clearly can be drawn in a plane.

(b) Can you think of a graph that is not planar? How would you prove that it is not planar? Proving the nonexistence of something is usually more difficult than proving its existence. This case is no exception. Intuitively, we would expect that spares graphs would be planar and dense graphs would be non-planar.

(c) The topic of planarity is a result of trying to restrict a graph to two dimensions. Is there an analogous topic for three dimesions? What graphs are indeed non-planar.

Answer to Note c: If a graph has only a finite number of vertices, it can always be drawn in three dimensions. This is not true for all graphs with an infinite number of vertices. The only "one-dimensional" graphs are the ones that consist of a finite number of chains, with one or more vertex in each chain.

Example:

A discussion of planarity is not complete without mentioning the famous Three Utilities Puzzle. The object of the puzzle is to supply three houses, A, B and C with the three utilities, gas, electric, and water. The constraint that makes this puzzle impossible to solve is that no utility lines may intersect.

There is a "solution" if you allow one of the lines to run under one house on its way to a second house. This solution is invalid when the puzzle is posed in the form of a graph, which is to draw the graph. So that no two edges intersect. This graph is one of the simplest non-planar graphs.

A planar graph divides the plane into one or more *regions*. Two points on the plane lie in the same region if you can draw a curve connected the two points that does not pass through an edge.

One of these regions will be of infinite area. Each point on the plane is either a vertex, a point on an edge, or a point in a region. A remarkable fact about the geography of planar graphs is the following theorem that is attributed to Euler.

SOLVED EXAMPLES

Example 1:

Consider the graph G given is Fig. 1.41 find the set V(G) of vertices of G and the set E(G) of edges of G. Find the degree of each vertex.

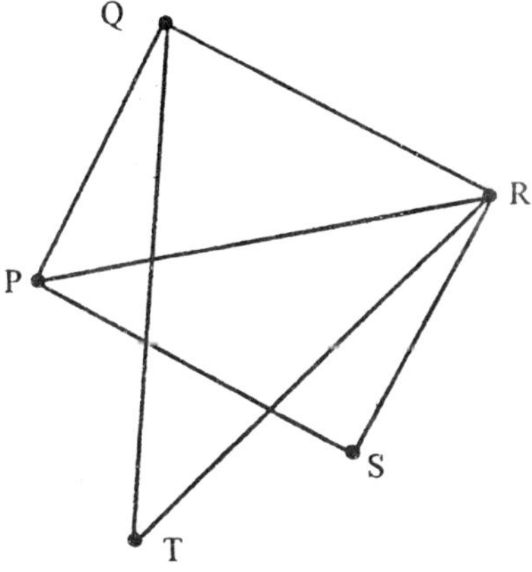

Fig. 1.41

Solution:

There are five vertices. So V(G) = {P, Q, R, S, T}.

There are seven pairs (a, b) of vertices where the vertex a is connected with the vertex b., hence E(G) = [{P, Q}, {P, R}, {P, S}, {Q, R}, {Q, T} {R, S}, {R, T}]

The degree of a vertex is equal to the number of edges to which it belongs. Since P belongs to three edges {P, Q}, {P, R} {P, S}. Hence

deg(P) = 3. Similarly, deg(Q) = 3,

deg(R) = 4, deg(S) = 2, deg(T) = 2.

Example 2:

Find the (i) vertex sets of components (ii) cut-vertices and (iii) cut-edges of the following graphs:

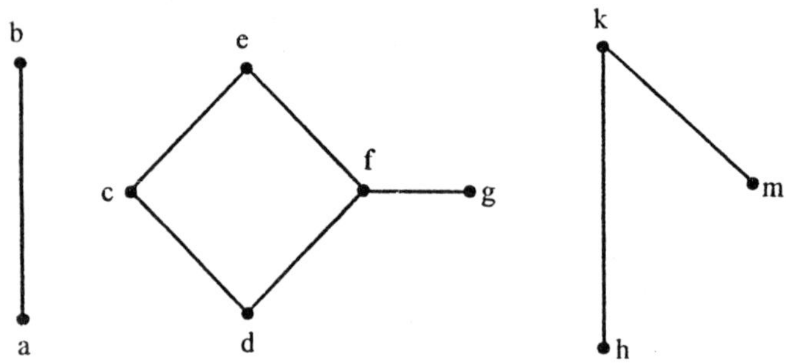

Fig. 1.42

Solution:

(i) The graph shown in Fig. 1.42 has three components. The vertex set of the components are {a, b}, {c, d, e, f, g} and {h, k, m}.

(ii) The cut-vertices of the graph are d and k.

(iii) Cut-edges of the graph are ab, cd, hk, and km.

Example 3:

In the graph shown in Fig. 1.43, find:

(i) a Hamiltonian circuit

(ii) a Hamiltonian circuit of minimal weight,

(iii) a Hamiltonian circuit of minimal weight beginning and ending at D.

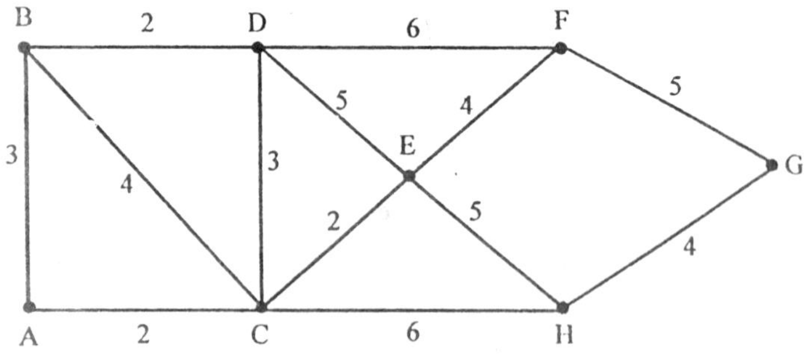

Fig. 1.43

Solution:

(i) A – B – D – F – G – H – E – C – A

(ii) C – A – B – D – F – G – H – E – C

(iii) D – B – A – C – E – H – G – F – D.

Example 4:

Verify that the graph G shown in Fig. 1.44 has an Eulerian cirucit.

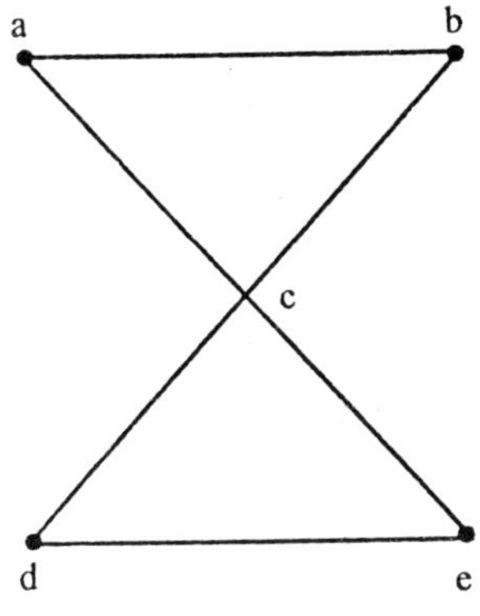

Fig. 1.44

Solution:

In the given graph G deg(a) = deg(b) = deg(c) = deg(d) = deg(e) = 2

i.e., all the vertices have even degree. Thus G has a Eulerian circuit. By inspection, the Eulerian circuit is as follows:

a – c – e – d – c – b – a.

Example 5:

The graphs in contain no Eulerian circuit.

Solution:

The graph does not contain Eulerian circuit since it is not connected.

Since all the vertices of the degree 3, hence it does not contain Eulerian circuit.

Example 6:

In the graph of Fig. 1.45, determine whether the following are paths, simple paths, trails, circuits or simple circuits.

(i) v_2

(ii) $v_5\ v_2\ v_3\ v_4\ v_4\ v_5$

(iii) $v_0\ e_1\ v_1\ e_{10}\ v_5\ e_9\ v_2\ e_2\ v_1$

(iv) $v_4\ e_7\ v_2\ e_9\ v_5\ e_{10}\ v_1\ e_3\ v_2\ e_9\ v_5.$

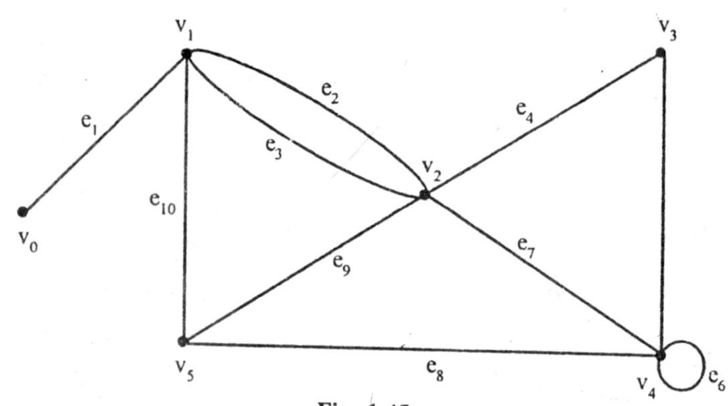

Fig. 1.45

Solution:

(i) It has no repeated edge, no repeated vertex, starts and ends at same vertex. Hence it is a simple circuit.

(ii) It has no repeated edge, no repeated vertex, starts and ends at same vertex, hence it is a circuit. Since vertex v_4 is repeated, hence it is not a simple circuit.

(iii) The given sequence has a repeated vertex v_1 but does not have a repeated edge, so it is a trail.

(iv) The given sequence has a repeated vertex v_2 and repeated edge e_9. Hence it is a path.

Example 7:

Which of the graphs Hamiltonian ciruciit. Give the circuits on the graphs that contain them.

Solution:

Contain Hamiltonian circuit since every circuit containing every vertex must contain the e_1 twice. But the graph does not have a Hamiltonian path $v_1 - v_2 - v_3 - v_4$.

Hamiltonian circuit given by $v_1\ e_1\ v_2\ e_2\ v_3\ e_3\ v_4\ e_4\ v_1$. Note that all vertices appear in this circuit but not all edges. The edge e_5 is not used in the circuit.

Example 8:

The two graphs are isomorphic.

Solution:

The correspondence between the two graphs is as follows :

(i) The vertices a, b, c, d and e correspond to v_1, v_2, v_3 and v_5 respectively.

(ii) The edges 1, 2, 3, 4, 5, and 6 correspond to e_1, e_2, e_3, e_4, e_5 and e_6 respectively.

Except for the names of their vertices and edges, isomorphic graphs are the same graph, perhaps drawn differently, two different ways of drawing the same graph, clearly they will be isomorphic graphs.

By the definition of isomorphism, it is clear that two graphs will be isomorphic if they have

(i) the same number of vertices,

(ii) the same number of edges.

(iii) an equal number of vertices with a given degree.

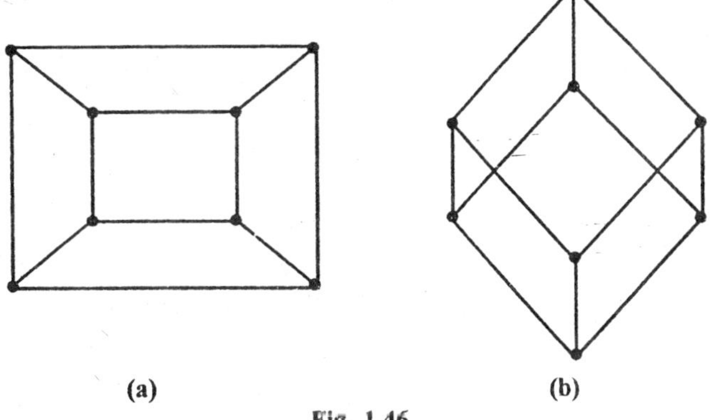

Fig. 1.46

However, these conditions are by no means sufficient. For example, the two graphs shown in Fig. 1.46 satisfy all three conditions, yet they are not isomorphic. If the graph in Fig. 1.46 (a) were to be isomorphic to the graph in Fig (b), vertex x must correspond to y, because there are no other vertices of degree 3.

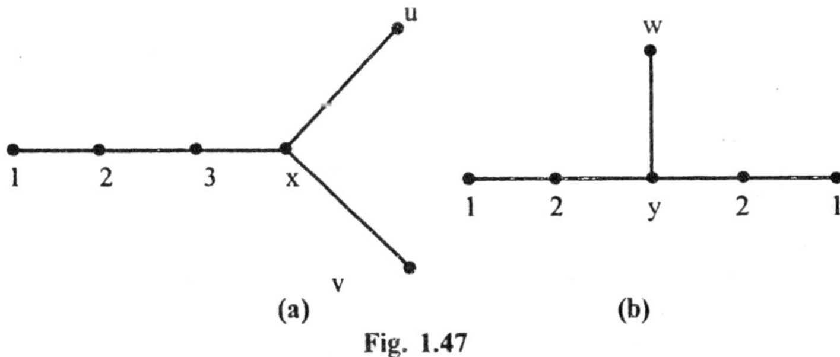

Fig. 1.47

Now in (b) there is only one pendsant vertex, w, adjacent to y, while in (a) there are two pendant vertices, u and v, adjacent to x.

Example 9:

Draw the graph G corresponding to the adjacency matrix

$$A(G) = \begin{matrix} 0 & 1 & 0 & 1 & 0 \\ 1 & 0 & 0 & 1 & 1 \\ 0 & 0 & 0 & 1 & 1 \\ 1 & 1 & 1 & 0 & 1 \\ 0 & 1 & 1 & 1 & 0 \end{matrix}$$

Solution:

Since A(G) is a 5 square matrix, hence G will have five vertices, say, v_1, v_2, v_3, v_4, v_5. Draw an edge from v_i to v_j when $a_{ij} = 1$. The graph is as shown in Fig. 1.48.

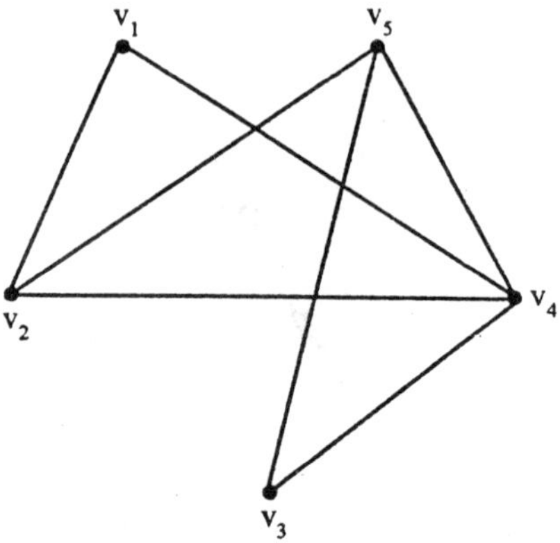

Fig. 1.48

EXERCISES

1. Draw a graph having the given properties or explain why no such graph exists:
   (a) Graph with six vertices each of degree 3.
   (b) Graph with five vertices of degree 0, 1, 2, 3.
   (c) Graph with four vertices of degree 1, 1, 2, 3.
   (d) Simple graph with four vertices of degree 1, 1, 3 and 3.

2. Determine which of the following sequences are paths,simple paths, cycle, and simple cycle.

   (a) $v_1\ e_8\ v_4\ e_3\ v_3\ e_7\ v_1\ e_8\ v_4$

   (b) $v_1\ e_1\ v_2\ e_6\ v_4\ e_3\ v_3\ e_2\ v_2$

   (c) $v_2\ e_2\ v_3\ e_3\ v_4\ e_4\ v_5\ e_5\ v_1\ e_1\ v_2$

   (d) $v_1\ e_1\ v_2\ e_2\ v_3\ e_4\ v_4\ e_4\ v_5$

3. Draw the graph represented by the following adjacency matrices:

(a)	0	1	0	1	0
	1	0	1	0	1
	0	1	0	1	1
	1	0	1	0	0
	0	1	1	0	0
(b)	0	0	0	0	1
	0	0	1	0	0
	1	0	0	0	0
	0	0	1	0	0
	0	1	0	0	0

4. Draw the graph represented by the incidence matrix.

v_1	1	0	0	0	0	1
v_2	0	1	1	0	1	0
v_3	1	0	0	1	0	0
v_4	0	1	0	1	0	0
v_5	0	0	1	0	1	1

2

GROUP THEORY AND APPLICATIONS

CYCLIC GROUPS

Groups are classified according to their size are structure. A group's structure is revealed by a study of it subgroups and other properties (e.g., whether it is abelian) that might give an overview of it. Cyclic groups have the simplest structure of all groups. In other words, the theory of groups, an important part in present day mathematics, started early in nineteenth century in connection with the solutions of algebraic equations, Originally a group was the set of all permutations of the roots of an algebraic equation which has the property that combination of any two of these permutations again belongs to the set. Later the idea was generalized to the concept of an abstract group. An abstract group is essentially the study of a set with an operation defined on it. Group theory has many useful applications both within and outside mathematics. Groups arise in a number of apparently unconnected subjects.

Definition: Cyclic Group, enerator. *Group G is cyclic if there exists a Î G such that the cyclic subgroup generated by a, (a), equals all of G. That is G = {na |n Î Z}, in which case a is called a generator of G. The reader should note that additive notation is used for G.*

Example 1:

*The group of additive integers is [**Z**; +] is cyclic:*

$$\boldsymbol{Z} = (1) = \{n * 1 \mid n \in \boldsymbol{Z}\}.$$

This observation does not mean that every integer is the product of an integer times 1. It means that

$$\boldsymbol{Z} = \{0\} \cup \underbrace{\{1 + \ldots + 1}_{\text{n terms}} | n \in N\} \cup \underbrace{\{(-1) + \ldots + (-1)}_{\text{n terms}} | -n \in N\}.$$

Example 2:

Z_{12} = [Z_{12} + $_{12}$], where + 12 is addition modulo 12, is a cyclic group. To prove this statement, all we need to do is demonstrate that some element of Z_{12} is a generator. One such element is 5; that is, (5) = Z_{12}. One more obvious generator is 1. In fact, 1 is a generator of every [$\mathbf{Z}_n$; + $_{12}$]. The reder is asked to prove that if an element is a generator, then its inverse is also a generator. Thus, –5 = 7 and –1 = 11 are the other generators of Z_{12}.

An example of "string art" that illustrates how 5 generates Z_{12}. Twelve tacks are placed alone a circle and numbered. A string is tied to tack 0, and is then looped around every fifth tack. As a result, the numbers of the tacks that are reached re exactly the ordered multiples of 5 modulo 12: 5, 10, 3, ... 7, 0. Note that if every seventh tack were used, the same artwork would be obtained. If every third tack were connected, the resulting loop would only use for tacks; thus 3 does not generate $\mathbf{Z}_{12}$.

PERMUTATION GROUPS

At the risk of boggling the reader's mind, we will now examine groups whose elements are functions. Recall that a *permutation on a set A* is a bijection from A into A. Suppose that A = {1, 2, 3}. There are 3 = 6 different permutations on A. They are listed in Table 1.1. The matrix form fro describing a function on a finite set is to list the domains across the top row and the image of each element directly below it. For example r_1 (1) = 2.

The operation that will give {i, r_1, r_2, f_1, f_2, f_3} a group structure is function composition. Consider the "product" r_1f_3

$$(r_1f_3)(1) = r_1(f_3(1)) = r_1(2) = 3$$

$$(r_1f_3)(2) = f_1(f_32)) = r_1(1) = 2$$

$$(f_1f_3)(3) = r_1(f_3(3)) = r_1(3) = 1$$

Table 2.1

Elements of S_3

$$i = \begin{pmatrix} 1 & 2 & 3 \\ 1 & 2 & 3 \end{pmatrix} \qquad f_1 = \begin{pmatrix} 1 & 2 & 3 \\ 1 & 3 & 2 \end{pmatrix}$$

$$r_1 = \begin{pmatrix} 1 & 2 & 3 \\ 2 & 3 & 1 \end{pmatrix} \qquad f_2 = \begin{pmatrix} 1 & 2 & 3 \\ 3 & 2 & 1 \end{pmatrix}$$

$$f_2 = \begin{pmatrix} 1 & 2 & 3 \\ 3 & 1 & 2 \end{pmatrix} \qquad f_3 = \begin{pmatrix} 1 & 2 & 3 \\ 2 & 1 & 3 \end{pmatrix}$$

Table 2.2

Table for S_3

	i	r_1	r_2	f_1	f_2	f_3
i	i	r_1	r_2	f_1	f_2	f_3
r_1	r_1	r_2	i	f_3	f_1	f_2
r_2	r_2	i	r_1	f_2	f_3	f_1
f_1	f_1	f_2	f_3	i	r_1	r_2
f_2	f_2	f_3	f_1	r_2	i	r_1
f_3	f_3	f_1	f_2	r_1	r_2	i

The images of 1, 2, and 3 under r_1f_3 and f_2 are identical. Thus, by the definition of equality for functions, we can say $r_1f_3 = f_2$. The complete table for the operation of function composition is given in Table 1.2. We don't even need the table to verify that we have a group.

(a) Function composition is associative

(b) The identity for the group is i. If g is any one of the permutations on A and $x \in A$.

$$gI(x) = g(i(x)) = g(x)$$

and $$ig(x) = i(g(x)) = g(x).$$

(c) If a permutation is displaced in matrix form, its inverse can be obtained by exchanging the two rows and rearranging the columns so that the top row is order. The first step is actually sufficient to obtain the inverse, but the sorting of the top row makes it easier to recognize the inverse.

Example:

If $$f = \begin{pmatrix} 1 & 2 & 3 & 4 & 5 \\ 5 & 3 & 2 & 1 & 4 \end{pmatrix},$$

$$f^{-1} = \begin{pmatrix} 5 & 3 & 2 & 1 & 4 \\ 1 & 2 & 3 & 4 & 5 \end{pmatrix}$$

$$= \begin{pmatrix} 1 & 2 & 3 & 4 & 5 \\ 4 & 3 & 2 & 5 & 1 \end{pmatrix}$$

Note from Table 1.2. that this group is non-abelian. Remember, non-abelian is the negation of abelian. The existence of two elements that don't commute is sufficient to make a group non-abelian. In this group, r1 and f3 is one such pair: $r_1f_3 \neq f_3r_1$.

NORMAL SUBGROUPS AND GROUP HOMOMORPHISMS

Our goal in this section is to answer an open question and introduce a related concept. The question is: When are left cosets of a subgroup a group under the induced operation? This question is open for non-abelian groups. Now that we have some examples to work with, we can try a few experiments.

NORMAL SUBGROUPS

Example:

$A_3 = \{i, r_1, r_2\}$ is a subgroup of S_3, an its left cosets are A_3 itself and $B_3 = \{f_1, f_2, f_3\}$. Whether $\{A_3, B_3\}$ is a group boils down to determining whether the induced operation is well defined. At this point we will ask for reader participation.

(a) Copy Table 15.3.2, the table for S_3.

(b) Shade in all occurrences of the elements of B_3 : f_1, f_2 and f_1. We will call these elements the gray elements of 3 the white ones.

Now consider the process of computing A_3B_3. The coset "product" is obtained by selecting one white element and one gray element. Note that white "time" gray is always gray. Thus, A_3B_3 is well defined. Similarly, the other three possible products are well defined. The table for the factor group S_3/A_3 is

	A_3	B_3
A_3	A_3	B_3
B_3	B_3	A_3

S_3/A_3 is just Z_2 in disguise. Note that A_3 and B_3 are also the right cosets of A_3.

GROUP

Definition : *An algebraic structure (G, o) where G is a non-empty set with a binary operation 'o' defined on it is said to be a group, if the binary operation satisfies the following axiom (called group axioms).*

(G_1) **Closure axiom :** G is closed under the operation o, i.e., $a \circ b \in G$, for all $a, b \in G$.

(G_2) **Associative axiom :** The binary operation o is associative i.e.,

$(a \circ b) \circ c = a \circ (b \circ c)\ \forall\ a, b, c \in G.$

(G_3) **Identity axiom :** There exists an element $e \in G$ such that

$e \circ a = a \circ e = a\ \forall\ a \in G.$

The element e is called the identity of 'o' in G.

(G_4) **Inverse axiom :** Each element of G possesses inverse, i.e. for each element $a \in G$, there exists an element $b \in G$ such that

The element b is then called the inverse of a with respect to 'o' and we write $b = a^{-1}$. Thus a^{-1} is an element of G such that

$$a^{-1} oa = a \text{ o } a^{-1} = e.$$

Abelian Group or Commutative Group Definition

A group (G, o) is said to be abelian or commutative if the composition 'o' is commutative, i.e., if,

$$a \text{ o } b = b \text{ o } a \; \forall \; a, b \in G.$$

A group which is not abelian is called non-abelian.

Example:

(i) The structures (**N**, +) and (**N**, ×) are not groups i.e., the set of natural numbers considered with the addition composition or me multiplication composition, does not form a group. For, the postulate (**G_3**) and (**G_4**) in the former case, and (**G_4**) in the latter case, are not satisfied.

(ii) The structure (**Z**, +) is a group, i.e., the set of integers with the addition composition is a group. This is so because addition in numbers is associative, the additive identity O belongs to **Z**, and the inverse of every element a, viz. –a belongs to **Z**. This is known as *additive group of integers.*

The structure (**Z**, ×), i.e. the set of integers with the multiplication composition does not form a group, as the axiom (G_4) is not satisfied.

(iii) The structures (**Q**, +), (**R**, +), (**C**, +) are all groups, i.e., the sets of rational numbers, real numbers, complex numbers, each with the additive composition, form a group.

But the same sets with the multiplication composition do not form a group, for the multiplicative inverse of the number zero does not exist in any of them.

(iv) The structure (**Q_0**, ×) is group, where **Q_0** is the set of non-zero rational numbers. This is so because the operation is associative, the multiplicative identity 1 belongs to **Q_0** and the multiplicative inverse of every element a in the set is 1/a, which also belongs to **Q_0**. This is known as the *multiplicative group* of non-zero rationals. Obviously (**R_0** ×) and (**C_0**, ×) are groups, where **R_0** and **Co**) are respectively the sets of non-zero real numbers and non-zero complex numbers.

(v) The structure (Q^+, ×) is a group, where Q^+ is the set of positive rational numbers: It can easily be seen that all the postulates of a group are satisfied.

Similarly, the structure (R^+, ×) is a group, where R^+ is the set of positive real numbers.

(vi) The, groups in (ii), (iii), (iv) and (v) above are all *abelian groups*, since addition and multiplication are both commutative operations is numbers.

Finite and Infinite Groups

If a group contains a finite number of distinct elements, it is called *finite group* otherwise an infinite group.

In other words, a group (G, o) is said to be finite or infinite according as the underlying set **G** is finite or infinite

Order of a Group

The number of elements in a finite group is called *order* of the group. An infinite group is said to be of *infinite order.*

Note : It should be noted that the smallest group for a given composition is the set {e} consisting of the identity element e alone.

COMPOSITION (OR OPERATION) TABLE

A binary operation in a finite set can completely be described by means of a table. This table is known as *composition table.* The composition table helps us to verify most of the properties satisfied by the binary operations.

This table can be formed as follows :

(i) Write the elements of the set (which are finite in number) in a row as well as in a column.

(ii) Write the element associated to the ordered pair (a_i, a_j) at the intersection of the row headed by a_i and the column headed by a_j. Thus

(ith entry on the left) . (jth entry on the top)

= (entry on the ith row and jth column intersect).

For example, the composition table for the group {(0, 1, 2, 3,4)} for the operation of addition is given below :

+	0	1	2	3	4
0	0	1	2	3	4
1	1	2	3	4	5
2	2	3	4	5	6
3	3	4	5	6	7
4	4	5	6	7	8

In the above example, the first element of the first row in the body of the table, 0 is obtained by adding the first element 0 of head row and the first element 0 of the head column. Similarly the third element of 4th row (5) is obtained by adding the third element 2 of the head row and the fourth element of the head column and so on.

An operation represented by the composition table will be binary, if every entry of the composition table belongs to the given set. It is to be noted that composition table contains all possible combinations of two elements of the set with respect to the operation.

Note :

(i) If should be noted that the elements of the set should be written in the same order both in top border and left border of the table, while preparing the composition table.

(ii) Generally a table which defines a binary operation '.' on a set is called *multiplication table*, when the operation is '+' the table is called *an addition table.*

Group Tables

The composition tables are useful in examining the following axioms in the manner explained below:

1. **Closure Property :** If all the elements of the table belong to the set **G** (say) then **G** is closed under the Composition o (say). If any of the elements of the table does not belong to the set, the set is not closed.
2. **Existence of Identity :** The element (in the vertical column) to the left of the row identical to the top row (border row) is called an identity element in the **G** with respect to operation 'o'.
3. **Existence of Inverse :** If we mark the identity elements in the table then the element at the top of the column passing through the identity element is the inverse of the element in the extreme left of the row pussing through the identity element and vice-versa.

4. **Commutativity :** It the table is such that the entries in every row coincide with the corresponding entries in the corresponding column i.e., the composition table is symmetrical about the principal or main diagonal, the composition is said to have satisfied the commutative axiom otherwise it is not commutative.

The process will be more clear with the help of following illustrative examples.

GENERAL PROPERTIES OF GROUPS

Theorem 1:

If corresponding to any element $a \in G$; there is an element which satisfies one of the conditions

$$a + 0_a = a \text{ or } 0_a + a = a$$

then it is necessary that $O_a = 0$, where 0 is the identity element of the group.

Proof:

Since 0 is the identity element,

We have

$$a + 0 = a \qquad \text{...(i)}$$

Also, it is given that

$$a + 0_a = a \qquad \text{...(ii)}$$

Hence, from (i) and (ii)

$$a + 0_a = a + 0$$

or $\quad 0_a = 0 \quad$ (by left cancellation law)

Again, we have

$$0 + a = a \qquad \text{...(iii)}$$

and $\quad 0_a + a = a \quad$ (given $\quad$ ...(iv)

Hence, from (iii) and (iv), we get

$$0_a + a = 0 + a$$

so that $0_a = 0 \quad$ (by right cancellation law.)

Theorem 2:

The inverse of the product of two elements of a group G is the product of the inverse taken in the reverse order i.e.,

$$(a \circ b)^{-1} = b^{-1} \circ a^{-1} \;\forall\; a, b \in G$$

Proof:

Let us suppose a and b are any two elements of G. If a^{-1} and b^{-1} are inverses of a and b respectively, then

$a^{-1} \text{ o } a = e = a \text{ o } a^{-1}$ (e being the identity element)

and $b^{-1} \text{ o } b = e = b \text{ o } b^{-1}$

Now, $(a \text{ o } b) \text{ o } (b^{-1} \text{ o } a^{-1}) = [(a \text{ o } b) \text{ o } b^{-1}] \text{ o } a^{-1}$ (by associativity)

$= [a \text{ o } (b \text{ o } b^{-1})] \text{ o } a^{-1}$ (by associativity)

$= [a \text{ o } (b \text{ o } b^{-1})] \text{ o } a^{-1}$ (by associativity)

$= (a \text{ o } e) \text{ o } a^{-1}$ $[\because b \text{ o } b^{-1} = e]$

$= a \text{ o } a^{-1}$ $[\because a \text{ o } e = a]$

$= e.$ $[\because a \text{ o } a^{-1} = e]$

Also $(b^{-1} \text{ o } a^{-1}) \text{ o } (a \text{ o } b) = b^{-1} \text{ o } [a^{-1} \text{ o } (a \text{ o } b)]$ (by associativity)

$= b^{-1} \text{ o } [(a^{-1} \text{ o } a) \text{ o } b]$

$= b^{-1} \text{ o } (e \text{ o } b)$ $[\because a^{-1} \text{ o } a = e]$

$= b^{-1} \text{ o } b$ $[\because e \text{ o } b = b]$

$= e.$

Hence, we have

$(b^{-1} \text{ o } a^{-1}) \text{ o } (a \text{ o } b) = e = (a \text{ o } b) \text{ o } (b^{-1} \text{ o } a^{-1})$

Therefore, by definition of inverse, we have

$(a \text{ o } b)^{-1} = b^{-1} \text{ o } a^{-1}$

This theorem can be generalised as :

if a, b, c, ... k, l, m $\in$ G, then

$(a \text{ o } b \text{ o } c \text{ o } \ldots k \text{ o } l \text{o } m)^{-1} = m^{-1} \text{ o } l^{-1} \text{ o } k^{-1} \text{ o} \ldots c^{-1} \text{ o } b^{-1} \text{ o } a^{-1}$

Theorem 3:

Cancellation laws hold good in a group, i.e., if a, b, c, are any elements of G, then

$a \text{ o } b = a \text{ o } c \Rightarrow b = c$ *(left cancellation law)*

and $b \text{ o } a = c \text{ o } a \Rightarrow b = c$ *(right cancellation law)*

Proof:

Let a $\in$ G. Then

$a \in G \Rightarrow G$ such that $a^{-1} \text{ o } a = e = a \text{ o } a^{-1}$,

where e is the identity element

Now, let us assume that

$a \circ b = a \circ c$

then $a \circ b = a \circ c$

$\Rightarrow a^{-1} \circ (a \circ b) = a^{-1} (a \circ c)$

$\Rightarrow (a^{-1} \circ a) \circ b = (a^{-1} \circ a) \circ c$ (by associative law)

$\Rightarrow e \circ b = e \circ c$ $(\because a^{-1} \circ a = e)$

$\Rightarrow b = c.$

Similarly, $b \circ a = c \circ a$

$\Rightarrow (b \circ a) \circ a^{-1} = (c \circ a) \circ a^{-1}$

$\Rightarrow b \circ (a \circ a^{-1}) = c \circ (a \circ a^{-1})$

$\Rightarrow b \circ e = c \circ e$

$\Rightarrow b = c.$

Theorem 4:

G is a group with binary operation o and if a and b are any elements of G, then the linear equations

$$a \circ x = b \text{ and } y \circ a = b$$

have unique solutions in G.

Proof:

Now $a \in G \Rightarrow a^{-1} \in G,$

and $a^{-1} \in G, b \in G \Rightarrow a^{-1} \circ b \in G.$

Substituting $a^{-1} \circ b$ for x in the equation $a \circ x = b$, we obtain

$a \circ (a^{-1} \circ b) = b$

$\Rightarrow (a \circ a^{-1}) \circ b = b$

$\Rightarrow e \circ b = b$

$b = b$ [$\because$ e is the identity]

Thus $x = a^{-1} ob$ is a solution of the equation $aox = b$.

To show that the solution is unique let us suppose that the equation $aox = b$ has two solutions given by

$$x = x_1 \text{ and } x = x_2$$

Then $aox_1 = b$ and $aox_2 = b$

$\Rightarrow aox_1 = aox_2 = b$

$\Rightarrow x_1 = x_2$ (by left cancellation law)

In a similar manner, we can prove that the equation

$$y o a = b$$

has the unique solution

$$y = b \text{ o } a^{-1}.$$

Theorem 5(a):

The identity element of a group is unique.

Proof:

Let us suppose e and e' are two identity elements of group **G**, with respect to operation o.

Then $\quad e \text{ o } e' = e$ if e' is identity.

and $\quad e \text{ o } e' = e'$ if e is identity.

But $\quad e \text{ o } e'$ is unique element of **G**, therefore,

$$e \text{ o } e' = e \text{ and } e \text{ o } e' = e' \Rightarrow e = e'$$

Hence the identity element in a group is unique.

Theorem 5(b):

If the inverse of a is a^{-1} then the inverse a^{-1} is a, i.e., $(a^{-1})^{-1} = a$.

Proof:

If e is the identity element, we have $a^{-1} \text{ o } a = e$ (by definition of inverse)

$\Rightarrow (a^{-1})^{-1} \text{ o } (a^{-1} \text{ o } a) = (a^{-1})^{-1} \text{ o } e \qquad [\because a^{-1} \in \mathbf{G} \Rightarrow (a^{-1})^{-1} \in G]$

$\Rightarrow [(a^{-1})^{-1} \text{ o } a^{-1}] \text{ o } a = (a^{-1})^{-1}$

[$\because$ Composition in G is associative and e is identity element]

$\Rightarrow e \text{ o } a = (a^{-1})^{-1}$

$\Rightarrow a = (a^{-1})^{-1}$

$\Rightarrow (a^{-1})^{-1} = a.$

Theorem 5(c):

*The inverse of each element of a group is unique, i.e., in a group **G** with operation o for every a $\in$ **G**, there is only one element a^{-1} such that*

$$a^{-1} \text{ o } a = a \text{ o } a^{-1} = e, \text{ e being the identity.}$$

Proof:

Let a be any element of a group **G** and let e be the identity element. Suppose there exist a^{-1} and a' two inverses of a in **G** then

$$a^{-1} o\ a = e = a\ o\ a^{-1}$$

and $$a'\ o\ a = e = a'\ o\ a$$

Now, we have

$a^{-1} o\ (a\ o\ a') = a^{-1} o\ e$ (since $a\ o\ a' = e$)

$= a^{-1}$ ($\because$ e is identity)

Also, $(a^{-1} o\ a)\ o\ a' = e\ o\ a'$ ($\because a^{-1} o\ a = e$)

$= a'$ ($\because$ e is identity)

But $a^{-1} o\ (a\ o\ a') = (a^{-1} o\ a)\ o\ a'$ as in a group composition is associative

$$\therefore\ a^{-1} = a^{1}$$

COSETS AND FACTOR GROUPS

Consider the group $[\mathbf{Z}_{12}\ ;\ +\ _{12}]$. As we saw in the previous section, we can picture its cycle properties with the string art here we will be interested in the non-generators, like 3. The solid lines the only one-third of the tacks have been reached by starting at zero and jumping to every third tack. The numbers of these tacks correspond to (3) = {0, 3, 6, 9}. What happens if you start at one of the unused tacks and again jump to every third tack? The two broken paths on that identical squares are produced. The tacks are thus partitioned into very similar subsets. The subsets of $\mathbf{Z}_{12}$ that they correspond to are {0, 3, 6, 9}. {1, 4, 7, 10} and {2, 5, 8, 11}. These subsets are called *cosets*. In particular, they are called *cosets of the subgroup* {0, 3, 6, 9}. We will see that under certain conditions, cosets of a subgroup can form a group of their own. Before pursuing this example any further we will examine the general situation.

Example:

In start at either 1 or 7 and obtain the same path by taking jumps of three tacks in each step. Thus

$$1 +_{12}\{0, 3, 6, 9\} = 7 +_{12}\{0, 3, 6, 9\} = \{1, 4, 7, 10\}.$$

The set of left (or right) cosets of a subgroup partition a group in a special way:

SOME DEFINITIONS

(i) If for every a $\in$ **G**, there exists an element e in **G**, such that eoa = a, then e is called, the *left identity*.

(ii) If for every $a \in$ **G**, there exists an element e in **G** such that aoe = a, then e is called the *right identity.*

(iii) If for an $a \in$ **G** there exists an element a^{-1} in **G** such that $a \circ a^{-1}$ = e, then a^{-1} is called the *left inverse* of a.

(iv) If for an $a \in G$ there exists an element a^{-1} in G such that $a \circ a^{-1}$ = e, then a^{-1} is called the *right inverse* of a.

Theorem 1:

The left inverse of an element is also its right inverse, i.e.,

$$a^{-1} \circ a = e = a \circ a^{-1}$$

Proof:

Now, $a^{-1} \circ (a \circ a^{-1}) = (a^{-1} \circ a) \circ a^{-1}$ (by associative law)

$= a^{-1} \circ e$ [by theorem 1]

Thus $a \circ (a \circ a^{-1}) = a^{-1} \circ e$

$\therefore$ $a \circ a^{-1} = e$ [by left cancellation law]

Hence a^{-1} is also the right inverse of a.

Theorem 2:

The left identity is also the right identity ie.,

$$e \circ a = a = a \circ e \ \forall \ a \in G$$

Proof:

If a^{-1} be the left inverse of a,

then $a^{-1} \circ (a \circ e) = (a^{-1} \circ a) \circ e$ (By associative law)

or $a^{-1} \circ (a \circ e) = e \circ e$ (by definition of left inverse)

$= e$

$= a^{-1} \circ a$

Thus $a^{-1} \circ (a \circ e) = a^{-1} \circ a$

$\therefore$ $a \circ e = a$ [by left cancellation law]

Hence e is also the right identity element.

An Alternative Definition for a Group

A set G with a binary composition denoted multiplicatively is a group if

(i) the composition is associative.

*(ii) for every pair of elements a, b $\in$ **G**, the equations $a x = b$ and $ya = b$ have unique solutions in G.*

Proof:

Binary operation implies that the set **G**, under consideration is closed under the operation. Now to prove set **G** a group we have to show that the left identity exists, and each element of **G** possesses left inverse with respect to the operation under consideration.

It is given that for every pair of elements a, b $\in$ **G** the equation ya = b has a solution in **G**. Therefore, if a $\in$ **G**, then taking b = a, we observe that there exists an element, say e $\in$ **G** such that

$$ea = a \qquad ...(i)$$

Now, let us suppose that b is any arbitrary element of **G**.

Therefore, there exists x $\in$ G such that

$$ax = b \qquad ...(ii)$$

Thus $\quad b = ax \Rightarrow eb = e\,(ax)$

$\Rightarrow \quad eb = (ea)\,x$ (associative law)

$\Rightarrow \quad eb = ax$ [from (i)]

$\Rightarrow \quad eb = b$ [from (ii)]

Therefore $\exists$ e $\in$ **G** such that

$eb = b \ \forall \ b \in$ **G**

$\therefore$e is the left identity,

Now, let b = e $\in$ **G** be an element.

$\therefore \quad ya = b \Rightarrow ya = e$

$\Rightarrow$ y is inverse of a in **G**.

Let, $y = a^{-1}$ such that

$a^{-1}\,a = e$ then $a^{-1} \in$ **G** as ya = e has got solution in **G**.

Thus, a^{-1} is the left inverse of a in **G**. Therefore each element of **G** possesses left inverse.

Hence, **G** is a group for the given composition if the postulates (i) and (ii) are satisfied.

Definition of a Group Based Upon Left Axioms

Let **G** *be a non-empty set equipped with a binary operation denoted by *, then this algebraic structure (*G, **) is a group if the binary operation * satisfies the following postulates :*

1. **Closure property, i.e.,** $ab \in G \ \forall \ a, b \in G$
2. **Associativity, i.e.,** $(ab)\, c = a\,(bc) \ \forall \ a, b, c \in G$
3. **Existence of Left Identity.** *There exists an element* $e \in G$ *such that* $ea = a \ \forall \ a \in G$. *The element is called the left identity.*
4. **Existence of Left Inverse.** *Each element of G possesses left inverse. In other words* $a \in G \Rightarrow$ *there exists an element* $a^{-1} \in G$ *such that* $a^{-1} a = e$. *The* element a^{-1} is the left inverse of a.

Proof:

This definition of a group and the classical definition of a group given is 14.1 are equivalent, obviously if the postulates of a group given is § 14.1. hold good, the postulates of a group given in this definition will also hold good.

If the postulate of a group given in this definition hold good, then the postulates given will also hold good if starting with left axioms we prove that the left identity is also the right Identity and the left inverse a element is also the right inverse.

MODULO SYSTEM

It is of common experience that railway time-table is fixed with the provision of 24 hours in a day and night. When we say that a particular train is arriving at 15 hours, it implies that the train will arrive at 3 p.m. according to our watch. Thus all the timing starting from 12 to 23 hours correspond to one of 0, 1, 3 ... 11 O'clock as indicated in watches. In other words all integers from 12 to 23 are equivalent to one or the other of integers 0, 1, 2, 3, ...11 with modulo 12. In saying like this the integers in question are divided into 12 classes.

In the manner described above the integer could be divided into 2 classes, or 5 classes or m (m being a positive integer) classes and then we would have written mod 2 or mod 5 or mod m. This system of representing integers is called *modulo system*.

Addition Modulo m

We shall now define a new type of addition known as "addition modulo m". and written as $a +_m b$ where a and b are any integers and m is a fixed positive integer.

By definition, we have

$$a +_m b = r, \ 0 \le r < m$$

where r is the least non-negative remainder when a + b, i.e., the ordinary sum of a and b, is divided by m.

For example $5 +_6 3 = 2$, since $5 + 3 = 8 = 1\ (6) + 2$, i.e., 2 is the least non-negative remainder when 5 + 3 is divided by 6. Similarly, $5 +_7 2 = 0$,

$$4 +_3 2 = 0;\ 3 +_3 1 = 1,\ 15 +_5 7 = 2.$$

Thus to find $a +_m b$, we add a and b in the ordinary way and then from the sum, we remove integral multiples of m in such a way that the remainder r is either 0 or a positive integer less man m.

When a and b are two integers such that a – b is divisible by a fixed positive integer m, then we write

$$a = b \pmod{m}$$

which is read as "a is congruent to b modulo m."

Thus $a = b \pmod{m}$ if $a - b$ is divisible by m. For example $13 = 3 \pmod{5}$ since $13 - 3 = 10$ is divisible by 5, $5 = 5 \pmod{5}$, $16 = 4 \pmod{6}$; $-20 = 4 \pmod{6}$.

Multiplication Modulo p

We shall now define a new type of multiplication known as "multiplication modulo p" and written as $a \times_p b$ where a and b are any integers and p is a fixed positive integer.

By definition we have

$$a \times_p b = r,\ 0 \le r \le p,$$

where r is the least non-negative remainder when ab, i.e., the ordinary product of a and b, is divided by p. For example $4 \times_7 2 = 1$, since $4 \times 2 = 8 = 1\ (7) + 1$.

It can be easily shown that if $a = b \pmod{p}$, men $a \times_p C = b \times_p C$

Additive Group of Integers Modulo m

The set G = (0, 1, 2, ... m –1} of first m non-negative integers is a group. The composition being addition reduced modulo m.

Closure Property : We have by definition of addition modulo m,

$$a + mb = r$$

where r is the lest non-negative remainder when the ordinary sum a + b is divided by m. Obviously $o \le r \le m - 1$. Therefore for all a b ∈ G we have a + mb ∈ G and thus G is closed with respect to the composition addition modulo m.

Associative Property : Let a, b, c be any arbitrary elements in G. Then

$(a + b) +_m c = (a +_m b) +_m c$ $\qquad [\because b +_m c = b + c \pmod{m}]$

= least non-negative remainder when a + (b + c) is divisible by m

= least non-negative remainder when (a + b) + c is divided by m, since

$$a + (b + c) = (a + b) + c$$

= (a + b) $+_m$ c [by definition of $^+$m]

= (a $+_m$ b) $+_m$ c [$\because$ a + b = a $+_m$ b (mod m)]

$\therefore$ '$+_m$' is an associative composition.

Existence of Identity Element : We have 0 ∈ **G**. Also, if a is any element of **G**, then 0 $+_m$ a = a = a $+_m$ 0. Therefore 0 is the identity element.

Existence of Inverse : The inverse of 0 is 0 itself. If r ∈ **G** and r ≠ 0, then

m – r ∈ **G**. Also (m – r) $+_m$ r = 0 = r $+_m$ (m – r). Therefore (m – r) is the inverse of r.

Commutative Property : The composition '+m' is commutative also.

Since

a $+_m$ b = least-non-negative remainder when a + b is divided by m

= least non-negative remainder when b + a is divided by m

= b $+_m$ a.

The set **G** contains" m elements.

Hence (**G**, $+_m$) is a finite abelian group of order m.

Multiplicative Group of Integers Modulo p where p is Prime

The set G of (p – 1) integers 1, 2, 3, ...p – 1, p being prime, is finite abelian group of order p – 1 the composition being multiplication modulo p.

Let G = (l, 2, 3, ...p – 1} where p is prime.

Closure Property : Let a and b be any elements of **G**. Then

$1 \le a \le p - 1$, $1 \le b \le p - 1$. Now by definition. a $\times_p$ b = r where r is the least non-negative remainder when the ordinary product a b is divided by p. Since p is prime, therefore a b is not exactly divisible by p. Therefore r can not be zero and we shall have $1 \le r \le p - 1$. Thus a $\times_p$ b ∈ **G** ∀ a, b ∈ **G**. Hence the closure axiom is satisfied.

Associative Law : Let a, b, c, be any arbitrary elements of **G**.

Then a $\times$ ($_p$b $\times_p$ c) = a $\times_p$ (bc) [$\because$ b $\times_p$ c = bc(mod p)]

= least non-negative remainder when a(bc) is divided by p

= least non-negative remainder when ab(c) is divided by p

= (ab) $\times_p$ c

$= (a \times_p b) \times_p c$ $\quad [\because ab = a \times_p b (\text{mod } p)]$

$\therefore$ '$\times_p$' is an associative composition.

Existence of left identity : We have $1 \in G$. Also if a is any element of G, then $1 \times_p a = a$. Therefore 1 is the left identity.

Existence of left inverse : Lets be any member of G.

Then $\quad 1 < s < p - 1$.

Let us consider the following (p – 1) products :

$1 \times_p s, 2 \times_p s,$

$3 \times_p s, \ldots (p - n) \times_p s.$

All these are elements of G. Also no two of these can be equal as shown below :

Let i and j be two unequal integer such that

$1 \leq i \leq p - 1, 1 \leq j \leq p - 1$ and $i > j$.

Then $\quad 1 \times_p s = j \times_p s$

$\Rightarrow$ is and js leave the same least non-negative remainder when divided by p

$\Rightarrow$ is -j s is divisible by p.

$\Rightarrow$ (i - j) s is divisible by p.

Since $1 \leq (i - j) < p - 1$; $1 \leq s \leq p - 1$ and p is prime, therefore (i – j) s can not be divided by p.

$\therefore \quad i \times_p s \neq j \times_p s.$

Thus $1 \times_p s, 2 \times_p s, \ldots (p - 1) \times_p s$ are (p – 1) distinct elements of the set G. Therefore one of these elements must be equal to 1.

Let Then $s' \times_p s = 1$. Then s' is the left inverse of s.

Commutative Law : The composition '$\times_p$' is commutative, since

$a \times_p b$ = least non-negative remainder when ab is divisible by p

= least non-negative remainder when ba is divided by p

$= b \times_p a$

$\therefore$ $(G, \times_p)$ is a finite abelian group of order p – 1.

Theorem 1:

The set of non-zero residue classes modulo p. Where p is a prime, forms a group with respect to multiplication of residue classes.

Proof:

Let **I** = {...., –3, –2, –1, 0, 1, 2, 3, ...} be the set of integers. Let $a \in$ **I**, then {a} is residue class modulo p of **I**, if {a} = {x : $x \in$ **I** and x – a is divisible by p}.

If p|a then {a} = {0} which is called the zero residue class. Let **G** be the set of non-zero residue classes mod p(p being prime) then

$$G = \{1, 2, 3, \ldots\ldots (p-1)\}$$

Closure axiom : Let $r_1, r_2 \in$ **G** then

$$r_1 \cdot r_2 = r \pmod{p}$$

where, r is the least non-negative integer such that $0 \le r \le p - 1$ obtained after dividing r_1, r_2 by p.

Also, since p is prime r_1, r_2 is not divisible by p. Hence r cannot be zero.

Hence, $r_1 \cdot r_2 = r = G$.

Thus closure axiom is satisfied

Associative axiom : Multiplication of residue classes is associative.

Existence of Identity : $I \in$ **G** and $a \cdot 1 = 1\, a = a \;\forall\; a \in$ **G**.

Therefore, 1 is the identity element in **G** with respect to multiplication.

Existence of Inverse. Let $s \in G$ then $1 \le s \le p - 1$. Let us consider following (p – 1) elements.

$$1 \cdot s, 2 \cdot s, 3 \cdot s \ldots, (p-1) \cdot s$$

All these elements are elements of **G** because the closure law is true. All these elements are distinct as otherwise if

$i.s = j.s$ for $i \neq j$ and $j, j \in$ **G**

then $i.s = j.s \Rightarrow i.s - j.s$ is divisible by p

$\Rightarrow$ (i – j).s is divisibly by p

$\Rightarrow$ (i – j) is divisibly by p $\quad [\because 1 \le s < p - 1]$

$\Rightarrow$ i = j which is contrary to our assumption that $i \neq j$,

Therefore above (p – 1) elements are the same as the elements of **G**. Hence some one of them should be 1 also. Let $s'.s = 1$ where $1 \le s' \le p - 1$. Hence s' is inverse of s. Hence inverse axiom is also satisfied.

$\therefore$ **G** is a group under multiplication mod p.

Note: Since $r.s = s.r \;\forall\; r, s \in$ **G**.

G is finite abelian group of order (p — 1).

Theorem 2:

The residue classes modulo form a finite group with respect to addition of residue classes.

Proof:

Let **G** be the set of residue classes (mod m), then

$$\{G = \{\{0\}, \{1\}, \ldots\{r_1\}, \ldots\{r_2\}, \ldots \{m-1\}\}$$

or $$G = \{0, 1, 2, \ldots\ r_1, \ldots\ r_2 \ldots\ m-1 \ (\text{mod } m)\}$$

Closure axiom : $\{r_1\} + \{r_2\} = \{r_1 + r_2\} = \{r\} \in G$,
where r is the least positive integer obtained as remainder when $r_1 + r_2$ is divided by m $(0 \leq r < m)$.

Thus, the closure axiom is satisfied.

Associative axiom : The addition is associative.

Identity axiom : $\{0\} \in (G)$ and $\{0\} + \{r\} = \{r\}$. Hence the identity for addition is $\{0\}$.

Inverse axiom : Since $\{m-r\} + \{r\} = \{m\} = \{0\}$, the additive inverse of the element (r) is $(m - r)$,

Hence **G** is a finite group with respect to addition modulo m.

PERMUTATIONS

Definitions : *Suppose S is a finite set having n distinct elements. Then a one-one mapping of S onto itself is called a permutation of degree n.*

The number of elements in the finite set S is known as the degree of permutation.

Symbol for a permutation : Let $S = \{a_1, a_2, a_3, \ldots, a_n\}$ be a finite set having n distinct element. If $f : S \rightarrow S$ is one-one onto mapping, then f is permutation of degree n.

Let $f(a_1) = b_1$, $f(a_1) = b_2$, $f(a_3) = b_3$, ... , $f(a_n) = b_n$,

where $\{b_1, b_2, \ldots, b_n\} = \{a_1, a_2, a_3, \ldots, a_n)$, i.e. $b_1, b_2, \ldots . b_n$ is one arrangement of the n elements $a_1, a_2, \ldots, a_n$.

It is customary to write a permutation in a two line symbol. In this notation we write.

$$f = \begin{pmatrix} a_1 & a_2 & a_3 & \ldots & a_n \\ b_1 & b_2 & b_3 & \ldots & b_n \end{pmatrix}$$

i.e., each element in the second row is the f-image of the element of the first row lying directly above it.

If S = {1, 2, 3, 4} be a finite set of order four then

$$f = \begin{pmatrix} 1 & 2 & 3 & 4 \\ 2 & 4 & 1 & 3 \end{pmatrix}, \; g = \begin{pmatrix} 1 & 2 & 3 & 4 \\ 1 & 3 & 2 & 4 \end{pmatrix},$$

$$h = \begin{pmatrix} 1 & 2 & 3 & 4 \\ 1 & 2 & 4 & 3 \end{pmatrix}$$

etc. are all permutations of degree four, here in the permutation f the elements 1, 2, 3, 4, have been replaced respectively by the elements 2, 4, 1, 3. Thus f (1) = 2, f (2) = 4, f (3) = 1, f (4) = 3.

Similarly g (1) = 1, g (2) = 3, g (3) = 2, g (4) = 4

and h (1) = 1, h (2) = 2, h (3) = 4, h (4) = 3.

Thus for a permutation f on S, we just put the elements of S in one row in any order we like and below each element of this row we put down its image under f, g, or h to obtain another row of elements of S.

Equality of Two Permutations

Two permutations f and g of degree n are said to be equal if we have $f(a) = g(a) \;\forall\, a \in S$.

Example:

If $f = \begin{pmatrix} 1 & 2 & 3 & 4 \\ 2 & 3 & 4 & 1 \end{pmatrix}$, and $g = \begin{pmatrix} 2 & 4 & 1 & 3 \\ 3 & 1 & 2 & 4 \end{pmatrix}$

are two permutations of degree 4, then we have f = g. Here we see that both f and g replace I by 2,2 by 3, 3 by 4 and 4 by 1.

If $f = \begin{pmatrix} a_1 & a_2 & a_3 & \dots & a_n \\ b_1 & b_2 & b_3 & \dots & b_n \end{pmatrix}$

is a permutation of degree n, we can write it in several ways. The interchange of columns will not change the permutation. Thus, we can write :

If $f = \begin{pmatrix} a_2 & a_3 & a_1 & \dots & a_n \\ b_2 & b_3 & b_a & \dots & b_n \end{pmatrix} = \begin{pmatrix} a_n & a_1 & a_{n-1} \\ b_n & b_1 & a_{n-1} \end{pmatrix}$

$$= \begin{pmatrix} a_{n-1} & a_n & a_{n-2} & \dots & a_1 \\ b_{n-1} & b_n & b_{n-2} & \dots & b_1 \end{pmatrix}$$

Therefore, if f and g are two permutations of the same elements/degree n then it is always possible to write g in such a way that the first row of g coincide with the second row of f.

For example, if

$$f = \begin{pmatrix} a_1 & a_2 & a_3 & a_4 \\ a_2 & a_4 & a_1 & a_3 \end{pmatrix} \text{ and } g = \begin{pmatrix} a_1 & a_2 & a_3 & a_4 \\ a_4 & a_3 & a_2 & a_1 \end{pmatrix}$$

the by interchanging the columns of g we can write

$$g = \begin{pmatrix} a_2 & a_4 & a_1 & a_3 \\ a_3 & a_1 & a_4 & a_2 \end{pmatrix}$$

Total number of distinct permutations of degree n : If S is a finite set having n distinct elements, then we shall have n! distinct arrangements of the elements of S. Therefore there will be n! distinct permutations of degree n. If P_n be the set consisting of all permutations of degree n then the set P_n will have n! distinct elements. This set P_n is called the *symmetric set of permutations* of degree n. Sometimes it is also denoted by S_n. Thus P_n = (f : f is a permutation of degree n}.

The set P_3 of all permutations of degree 3 will have 3! i.e., 6 elements. Obviously

$$P_3 = \left\{ \begin{pmatrix} 1 & 2 & 3 \\ 1 & 2 & 3 \end{pmatrix}, \begin{pmatrix} 1 & 2 & 3 \\ 3 & 1 & 2 \end{pmatrix}, \begin{pmatrix} 1 & 2 & 3 \\ 2 & 3 & 1 \end{pmatrix}, \begin{pmatrix} 1 & 2 & 3 \\ 3 & 2 & 1 \end{pmatrix}, \begin{pmatrix} 1 & 2 & 3 \\ 1 & 3 & 2 \end{pmatrix}, \begin{pmatrix} 1 & 2 & 3 \\ 2 & 1 & 3 \end{pmatrix} \right\}$$

Identity Permutation : *If f is a permutation of degree n such that I replaces each element by the element itself, I is called the identity permutation of degree n.*

Thus $$I = \begin{pmatrix} 1 & 2 & 3 & \dots & n \\ 1 & 2 & 3 & \dots & n \end{pmatrix}$$

or $$\begin{pmatrix} a_1 & a_2 & a_3 & \dots & a_n \\ a_1 & a_2 & a_3 & \dots & a_n \end{pmatrix}$$

or $$\begin{pmatrix} b_1 & b_2 & b_3 & \dots & b_n \\ b_1 & b_2 & b_3 & \dots & b_n \end{pmatrix}$$

is the identity permutation of degree n.

Product or Composite of two permutations : *The Product or composite of two permutations f and g of degree n denoted by fg, is obtained by first carrying out the operation defined by f then by f then by g.*

Let us suppose P_n is the set of all permutations of degree n. Let

$$f = \begin{pmatrix} a_1 & a_2 & a_3 & \dots & a_n \\ b_1 & b_2 & b_3 & \dots & b_n \end{pmatrix}$$

and $$g = \begin{pmatrix} b_1 & b_2 & b_3 & \dots & b_n \\ c_1 & c_2 & c_3 & \dots & c_n \end{pmatrix}$$

be any two elements of P_n.

Here the permutation g has been written in such a way that the first row of g coincides with the second row of f. If the product of the permutations f and g is denoted multiplicatively, i.e., by fg, then by definition

$$fg = \begin{pmatrix} a_1 & a_2 & a_3 & \dots & a_n \\ c_1 & c_2 & c_3 & \dots & c_n \end{pmatrix}$$

For, f replaces a_1 by b_1 and then g replaces b_1 by c_1 so that fg replaces a_1 by c_1. Similarly fg replaces a_2 by c_2 a_3 by c_3 ,......, a_n by c_n,

Obviously, f g is also a permutation of degree n. Thus the product of two permutations of degree n is also a permutation of degree n. Therefore $fg \in P_n \ \forall \ f, g \in P_n$.

Inverse of permutations : If f be a permutation of degree n, defined on a finite set S consisting of a n distinct elements, then by definition f is a one-one mapping of S onto itself. Since f is one-one onto, it is invertible. Let f^{-1} be the inverse of map f then f^{-1} will be one-one onto map of S onto itself. Thus, f^{-1} is also a permutation of degree n on S. *This f^{-1} is known as the inverse of the permutations f.*

Thus, if $$f = \begin{pmatrix} a_1 & a_2 & a_3 & \dots & a_n \\ b_1 & b_2 & b_3 & \dots & b_n \end{pmatrix}$$

then $$f^{-1} = \begin{pmatrix} b_1 & b_2 & b_3 & \dots & b_n \\ a_1 & a_2 & a_3 & \dots & a_n \end{pmatrix}$$

Note : Evidently f^{-1} is obtained by interchanging the rows of f because

$$f(a_1) = b_1 \Rightarrow f^{-1}(b_1) = a_1 \text{ etc.}$$

GROUP OF PERMUTATIONS

The set P_n of all permutations on n symbols is a finite group of order n! with respect to composite of mappings as the operation. For $n \leq 2$, this group is abelian and for $n > 2$ it is always non-abelian.

Let $S = [a_1, a_2, a_3, ..., a_n]$ be a finite set having n distinct elements. Thus there are n! permutations possible on S. If P_n denotes the set of all permutations of degree n then multiplication of permutations on P_n satisfies the following axioms.

Closure axiom : Let $f, g \in P_n$ then each of them is one-one mapping of S onto itself and therefore their composite mapping (gof) is a one-one mapping of S onto itself. Thus (gof) is a permutation of degree n on S, i.e.,

$$f, g \in P_n \Rightarrow f g \in P_n.$$

This shows that P_n is closed under multiplication.

Associative axiom : Since the product of two permutations on a set S is nothing but the product of two one-one onto mappings on S and the product of mapping being associative, the product of permutations also obeys the associative law. Hence

$$(f\,g)\,h = f\,(gh) \text{ for } f, g, h \in P_n.$$

For example let

$$f = \begin{pmatrix} a_1 & a_2 & \dots & a_n \\ b_1 & b_2 & \dots & b_n \end{pmatrix}$$

$$g = \begin{pmatrix} b_1 & b_2 & \dots & b_n \\ c_1 & c_2 & \dots & c_n \end{pmatrix},\ h = \begin{pmatrix} c_1 & c_2 & \dots & c_n \\ d_1 & d_2 & \dots & d_n \end{pmatrix}$$

where $b_1, b_2 \dots, b_n$; $c_1, c_2, \dots, c_n$, and $d_1, d_2, \dots, d_n$ are simply different arrangements of the same n elements $a_1, a_2, \dots, a_n$.

Now, $$fg = \begin{pmatrix} a_1 & a_2 & \dots & a_n \\ c_1 & c_2 & \dots & c_n \end{pmatrix}$$

$$(fg)\,h = \begin{pmatrix} a_1 & a_2 & \dots & a_n \\ c_1 & c_2 & \dots & c_n \end{pmatrix}\begin{pmatrix} c_1 & c_2 & \dots & c_n \\ d_1 & d_2 & \dots & d_n \end{pmatrix}$$

$$= \begin{pmatrix} a_1 & a_2 & \dots & a_n \\ d_1 & d_2 & \dots & d_n \end{pmatrix}$$

Also, $$gh = \begin{pmatrix} b_1 & b_2 & \dots & b_n \\ c_1 & c_2 & \dots & c_n \end{pmatrix}\begin{pmatrix} c_1 & c_2 & \dots & c_n \\ d_1 & d_2 & \dots & d_n \end{pmatrix}$$

$$= \begin{pmatrix} b_1 & b_2 & \dots & b_n \\ d_1 & d_2 & \dots & d_n \end{pmatrix}$$

$\therefore$ f (gh) $$= \begin{pmatrix} a_1 & a_2 & \dots & a_n \\ b_1 & b_2 & \dots & b_n \end{pmatrix}\begin{pmatrix} b_1 & b_2 & \dots & b_n \\ d_1 & d_2 & \dots & d_n \end{pmatrix}$$

$$= \begin{pmatrix} a_1 & a_2 & \dots & a_n \\ d_1 & d_2 & \dots & d_n \end{pmatrix} = (fg)\,h.$$

Identity axiom : Identity permutation $I \in P_n$ is identity of multiplication in P_n because $If = fI = f\ \forall\ f \in P_n$.

Inverse axiom : Let $f \in P_n$ then f is one-one onto mapping, hence it is invertible. Hence f^{-1}, the inverse mapping of f is also one-one and onto. Consequently, f^{-1} is also a permutation in P_n.

$\therefore$ $$f^{-1} f = f f^{-1} = I.$$

For example let $f = \begin{pmatrix} a_1 & a_2 & \dots & a_n \\ b_1 & b_2 & \dots & b_n \end{pmatrix}$ be any element of P_n then

$$f^{-1} = \begin{pmatrix} b_1 & b_2 & \dots & b_n \\ a_1 & a_2 & \dots & a_n \end{pmatrix}$$

Now, $$f^{-1} f = \begin{pmatrix} b_1 & b_2 & \dots & b_n \\ a_1 & a_2 & \dots & a_n \end{pmatrix}\begin{pmatrix} a_1 & a_2 & \dots & a_n \\ b_1 & b_2 & \dots & b_n \end{pmatrix}$$

$$= \begin{pmatrix} b_1 & b_2 & \dots & b_n \\ b_1 & b_2 & \dots & b_n \end{pmatrix} = I$$

Similarly $$f\,f^{-1} = \begin{pmatrix} a_1 & a_2 & \dots & a_n \\ b_1 & b_2 & \dots & b_n \end{pmatrix} \begin{pmatrix} b_1 & b_2 & \dots & b_n \\ a_1 & a_2 & \dots & a_n \end{pmatrix}$$

$$= \begin{pmatrix} a_1 & a_2 & \dots & a_n \\ a_1 & a_2 & \dots & a_n \end{pmatrix} = I$$

Therefore f^{-1} is the multiplicative inverse of f.

Thus the symmetric set P_n of all permutations of degree n defined on a finite set forms a finite group of order n! with respect to the composite of permutations as the composition.

Commutative axiom : If we consider the symmetric group (P_1, o) of permutations of degree 1 with respect to permutation product o, then it consists of a single permutation namely the identity permutation I. Since I o I = I, (P_1, o) is an abelian group. If we consider the symmetric group (P_2, o) of all permutations of degree 2, i.e., the group of all permutations defined on a set of two elements (a_1, a_2), then

$$P_2 \left\{ \begin{pmatrix} a_1 & a_2 \\ a_1 & a_2 \end{pmatrix}, \begin{pmatrix} a_1 & a_2 \\ a_2 & a_1 \end{pmatrix} \right\}$$

Now $$\begin{pmatrix} a_1 & a_2 \\ a_1 & a_2 \end{pmatrix} o \begin{pmatrix} a_1 & a_2 \\ a_2 & a_1 \end{pmatrix} = \begin{pmatrix} a_1 & a_2 \\ a_2 & a_1 \end{pmatrix}$$

and $$\begin{pmatrix} a_1 & a_2 \\ a_2 & a_1 \end{pmatrix} o \begin{pmatrix} a_1 & a_2 \\ a_1 & a_2 \end{pmatrix} = \begin{pmatrix} a_1 & a_2 \\ a_2 & a_1 \end{pmatrix} \begin{pmatrix} a_1 & a_2 \\ a_2 & a_1 \end{pmatrix}$$

$$= \begin{pmatrix} a_1 & a_2 \\ a_2 & a_1 \end{pmatrix}$$

Therefore operation having commutative (P_2, o) is abelian group of order 2.

But when n > 2 then permutation product is not necessarily commutative. Hence (P_n, o) then is not necessarily an abelian group.

CYCLIC PERMUTATIONS

A permutation of the type

$$\begin{pmatrix} a_1 & a_2 & a_3 & \dots\dots & a_{n-1} & a_n \\ a_2 & a_3 & a_4 & \dots\dots & a_n & a_1 \end{pmatrix}$$ is called a cyclic permutation or a cycle. It is usually denoted by the symbol $(a_1, a_2, \dots a_n)$.

Thus if f is a permutation of degree n on a set S having n distinct elements and if it is possible to arrange some of the elements (say m in number) of the set S in a row such that the f-image of each element in this row is the element following it and the f-image of the last element in the row is the first element and the remaining (n – m) elements of the set S remain invariant under f, then f is called a cyclic permutation or a cycle of length m.

The number of objects permuted by the cycle is called the *length of cycle.*

Thus by the cycle of length one we mean a permutation in which the image of each element remains unchanged under a permutation f. Consequently, the cycle permutation of length one is the identity permutation.

One row symbol : One row symbol is used to denote a cyclic permutation. In this notation the elements of S are arranged in such a way that the image of each element in this row is the element which follows it and that of the last element is the first element. Those elements of **X** which remain invariant need not be written in the row.

Let $f = \begin{pmatrix} 1 & 2 & 3 & 4 & 5 & 6 \\ 2 & 4 & 1 & 3 & 5 & 6 \end{pmatrix}$ be a cyclic permutation.

Since the elements 1, 2, 3, 4 are such that f (1) = 2, f (2) = 4, f (4) and f (3) = 1 and two remaining elements 5 and 6 remain invariant under f, f is a cycle of length 4 or a 4-cycle and can be expressed as f(1 2 4 3).

Transposition : *A cycle of length two is called a transposition.* Thus the cycle (1, 3) is a transposition. It is a 2-cycle such that the image of 1 is 3 and image of 3 is 1 and the remaining missing *elements* are invariant.

Disjoint cycles : Two cycles are said to be disjoint if when expressed in one row notations, they have no element in common.

Some Theorems

Theorem 1:

Every permutation can be expressed as a product of transpositions.

Proof:

To prove the above result, we shall first show that every cycle can be expressed as a composite of transpositions. Let us consider a cycle $(a_1, a_2, ..., a_n)$ then

$$(a_1, a_2, ..., a_n) = (a_i\ a_n)\ (a_1\ a_{n-1})\ ...\ (a_1, a_2).$$

We have already proved that every permutation can be expressed as a composition of disjoint cycles. Therefore in the light of the two results stated above every permutation can be expressed as a product of transpositions.

Theorem 2(a):

The product of disjoint cycles is commutative.

Proof:

Let f and g be any two disjoint cycles, i.e., there is no element common in two when they are expressed in one row notation. Therefore, the elements permuted by f are invariant under g and vice-versa.

Hence, fog = gof; i.e., the product of disjoint cycles is commutative.

Theorem 2(b):

Every permutation can be expressed as a composite of disjoint cycles.

Proof:

Let the given permutation f be denoted by the usual two row symbol with in a bracket. Let a be any element in the first row and b the element in second row exactly beneath a, i.e., f (a) = b. Similarly, let f (b) = c. Continuing this process, an element 1 may be found in the upper row such that its f-image is a. Then (a, b, c, ...I) is one circular permutation. If there are additional elements a', b' etc., in the original permutation f, follow the above process to obtain another cycle, (a', b', c',...., I').

Even now if, some element or elements are left in original permutation this procedure can be repeated to the extent that all the elements of f are exhausted. In this way the original permutation can be put as the product of disjoint cycles.

EVEN AND ODD PERMUTATIONS

A permutation is said to be an even permutation if it can be expressed as a product of an even number of transpositions, otherwise, it is said to be an odd permutation.

Theorem 1:

Of the n! permutations on n symbols n/2! are even permutations and n/2! are odd permutations.

Proof:

Let the even permutation be $e_1, e_2, ..., e_m$ and the odd permutations be $O_1, O_2, ..., O_k$.

Then $$m + k = n!$$

Now let t be any transposition. Since t is evidently an odd permutation, we see that $te_1, te_2, ..., te_m$ are odd permutations and that $tO_1, tO_2, ..., tO_k$

are even permutations. Since an odd permutation is never an even permutation, we have

$$t\,e_1 \neq tO_j$$

for any i = 1, 2,... , m ; j = 1, 2, ... , k. Furthermore, if

$$t\,e_i = tej,$$

then $e_i = e_j$ by cancellation law.

Similarly, $tO_i, \neq tO_j$ if $i \neq j$

It follows that all the m even permutations must appear in the list tO_1, tO_2,..., tO_k, which are all distinct so that their number is m.

Similarly, all of the k odd permutations must be in the list.

$$te_1, te_2, ..., te_m.$$

which are all distinct as shown above and their number is k.

Hence, $m = k = 1/2n!$

Note:

1. A cycle containing an odd number of symbols is an even permutation, whereas a cycle containing an even number of symbols is an odd permutation, since a permutation on n symbols can be expressed as a product of (n – 1) transpositions.
2. Inverse of an even permutation is an even permutation and the inverse of an odd permutation is an odd permutation.
3. Product of two permutations is an even permutation if either both the permutations are even or both are odd and the product is an odd permutation if one permutation is odd and the other even.

Theorem 2:

A permutation can not be both even and odd, i.e., if a permutation f is expressed as a product of transpositions then the number of transpositions is either always even or always odd.

Proof:

Let us consider the polynomial A in distinct symbol $x_1, x_2, ... x_n$. It is defined as the product of $\frac{1}{2}n(n-1)$ factors of the form $x_i - x_j$ where $i < j$.

Thus
$$A = \prod_{i \leq j=1}^{n} (x_i - x_j)$$

$$= (x_1 - x_2)(x_1 - x_3)(x_1 - x_4) ... (x_1 - x_n)$$
$$(x_2 - x_3)(x_2 - x_4) ... (x_2 - x_n)$$

$$(x_3 - x_4) \ldots (x_3 - x_n) \ldots (x_{n-1} - x_n)$$

Consider now any permutation **P** on n symbols 1, 2, 3,...... n. By AP we mean the polynomial obtained by permutation the subscripts 1, 2...., n of the x_i as prescribed by **P**.

For example, taking n = 4, we have

$$A = (x_1 - x_2)(x_1 - x_3)(x_1 - x_4)(x_2 - x_3)(x_2 - x_4)(x_3 - x_4)$$

and if P = (1 3 4 2), then

$$\mathbf{AP} = (x_3 - x_1)(x_3 - x_4)(x_3 - x_2)(x_1 - x_4)(x_1 - x_2)(x_4 - x_3)$$

In particular if **P** = (l, 2) we have

$$\mathbf{AP} = (x_2 - x_1)(x_2 - x_3)(x_2 - x_4)(x_1 - x_3)(x_1 - x_4)(x_3 - x_4)$$
$$= -\mathbf{A}.$$

This shows that the effect of a transposition on A is to change the sign of A.

In general a transposition (i, j) i < j has the following effects on A:

(i) Any factor which involves neither the suffix i nor j remains unchanged.

(ii) The single factor $(x_i - x_j)$ changes its sign.

(iii) The remaining factors which involve either the suffix i or j but not both can be grouped into pairs of products, $\pm (x_m - x_j)(x_m - x_i)$ where $m \neq$ i or j and such a product remains unaltered when x_1 and x_j are interchanged.

Hence the net effect of transposition (i, j) on A is change its sign, i.e., A operated upon transposition (i, j) gives –A.

Now the permutation P considered as a product of s transpositions when operated upon A gives $(-1)^s$ A so that AP = $(-1)^s$ A and considered as a product of t transpositions gives $(-1)^t$ A so that AP = $(-1)^t$ A.

Hence $\qquad (-1)^s A = (-1)^t A$

or $\qquad (-1)^s = (1)^t$

Now, this equation will hold only if s and t are either both even or both odd. Hence the theorem.

ORBIT OF PERMUTATIONS

Let f be a permutation on a set S. If a relation ~ is defined on S such that $a \sim b \Leftrightarrow f^{(n)}(a) = b$ for some integer n $\forall$ a, b $\in$ S, we observe that the relation is

(i) Reflexive, because $a \sim a \Leftrightarrow f^{(o)}(a)$

$= I(a) = a \ \forall \ a \in S.$

(ii) Symmetric, because $a \sim b \Rightarrow f^{(n)}(a)$

$\Rightarrow$ b for some integer n

$\Rightarrow$ $a = f^{-(n)}(b)$

$\Rightarrow$ $b \sim a$ for $a, b \in S$.

(iii) Transitive, because $a \sim b$ and $b \sim c$

$\Rightarrow$ $f^{(n)}(a) = b$, $f^{(m)}(b) = c$ for some integers n and m.

$\Rightarrow$ $f^{(m)}(f^{(n)}(a)) = f^{(m)}(b) = c$

$\Rightarrow$ $f^{m+n}(a) = c$ for some integer $(m + n)$

$\Rightarrow$ $a \sim c$.

Thus, above defined relation $\sim$ is an equivalence relation on S and hence partitions it into mutually disjoint classes. Each equivalence class determined by the above relation is called an orbit of f.

INTEGRAL POWERS OF AN ELEMENT

Suppose **G** is a group and the composition has been denoted multiplicatively. Let $a \in$ **G**. Then by closure property a, aa, aaa, aaaa, etc., are all elements of **G**. Since the composition in **G** obeys general associative law, therefore a a a.... a to n factors is independent of the manner in which the factors may be grouped.

If n is a positive integer, we define a^n = a a a... a to n factors. Obviously $a^n \in G$.

If e is the identity element of the group **G**, then we define

$$a^0 = e.$$

If n is a positive integer then $-n$ is a negative integer. Now we define $a^{-n} = (a^n)^{-1}$ where $(a^n)^{-1}$ is the inverse of a^n in G. Thus $a^{-n} \in G$.

Thus we have defined for all integral values of n positive, zero or negative.

Integral Multiples of an Element of a Group

If in a group G the composition has been denoted additively, then in place of using the word Integral powers of an element of a group we use the word integral multiplies of an element of a group. The difference is only of notation otherwise the meaning is the same. Thus in this case if n is a positive integer we write n . a in place of a^n and we define $n\,a = a + a + ... + a$ upto n terms.

In place of a^0 we write 0a. Thus w define 0a = e where e is the identity of G.

If n is a positive integer, then in place of a^{-1} we write (– n) a. Thus we define (– n) a = – (n a) denotes the inverse of n a in **G**.

In multiplicative notation the following laws of indices can be easily proved:

$$a^m a^n = a^{m+n},$$

and $$(a^m)^n = a^{mn},$$

$\forall \in G$ and $\forall$ m ; n $\in$ I where I is the set of integers.

In additive notation the following laws of multiples can be easily proved;

$$ma + na = (m + n)a,$$

$$n\,(m\,a) = (n\,m)\,a,$$

$\forall$ a $\in$ **G** and $\forall$ m, $\in$ I.

Order of an Element of a Group

Definition: *If G is a group and a $\in$ G , the order (or period) of a is the least positive integer n such that*

$$a^n = e.$$

If there exists no such integer, we say that a is of infinite order or of zero order.

We shall use the notation o (a) for the order of a.

Note that the only element of order one in a group is the identity element e.

Important note: If there exists a positive integer m such that $a^m = e$, then the order of a is definitely finite. Also we must have o (a) $\leq$. m. When $a^m = e$, then the question of order of a being greater than m does not arise. At the most it can be equal to m. If m itself is the least positive integer such that $a^m = e$; then we will have

$$o\,(a) = m.$$

Some Theorems

Theorem 1:

The order of the elements a and x^{-1} ax are the same where a, x are any two elements of a group.

Proof:

Let n and m be the orders of a and x^{-1} ax respectively

Now $(x^{-1}ax)^2 = (x^{-1}a\,ax)\,(x^{-1}a\,x)$

$= x^{-1}a\,(xx^{-1})\,a\,x$

$= x^{-1} (a\, e)\, a\, x = x^{1}\, a\, a\, x = x^{-1} a^{2}x$.

In general, we get

$(x^{-1}ax)^{n} = x^{-1}\, a^{n}\, x = x^{-1}ex$ $\quad [\because\ o\,(a) = n \Rightarrow a^{n} = e]$

$= x^{-1}\, x = e.$

$\therefore \quad o(x^{-1}\, ax) \le n \Rightarrow m \le n.$

Again $\quad o(x^{-1}\, ax) \Rightarrow n = (x^{-1}\, ax) = e$

$\Rightarrow \quad x^{-1}\, a^{m}\, x = e$

$\Rightarrow \quad x^{-1}\, a^{m}\, x = x^{-1}\, x$

$\Rightarrow \quad a^{m}x = x$ (by left cancellation law)

$\Rightarrow \quad a^{m}x = ex$

$\Rightarrow \quad a^{m} = e$ (by right cancellation law)

$\Rightarrow \quad o(a) \le m \Rightarrow n \le m.$

Finally $m \le n,\ n \le m \Rightarrow m = n.$

Cor. *: Order of a b is the same as that of ba where a and b are any elements of a group.*

Proof:

We have $\quad a^{-1}\,(ab)a = (a^{-1}a)\,(ba)$

$= e(ba) = (ba)$

Thus, $\quad ba = a^{-1}(ab)a$

$\Rightarrow$ order of ba = order of $a^{-1}(ab)a$

$\Rightarrow$ order of ba = order ab $\quad [\because\ o(x^{-1}\, ax) = o(a)]$

Theorem 2(a):

If a is an element of order n and p is prime to n, then a^{p} is also of order n.

Proof:

Let m be the order of a^{p}.

Now, $\quad o(a) = n \Rightarrow a^{n} = e$

$\Rightarrow \quad (a^{n})^{p} = e^{p} = e$

$\Rightarrow \quad (a^{p})^{n} = e \Rightarrow\ o\,(a^{p}) \le e$

$\Rightarrow \quad m < n.$

Since p, n are relative primes, there exist integers x and y such that

$p_{x} + n_{y} = 1$

$\therefore$ $a = a^1 = a^{px+ny} = a^{px} \,.\, a^{ny} = a^{px}\,(a^n)^y$

$a^{px} \,.\, e^y = a^{px} \,.\, e = a^{px} = (a^p)^x$

Now, $a^m = [(a^p)^x]^m = (a^p)^{mx} = [a^p)^m]^x$

$= e^x$ $[\because o(a^p) = m]$

$= e$ $\Rightarrow (a^p)^m = e]$

$o\,(a) \leq m \Rightarrow n \leq m$.

Finally $m \leq n$ and $n \leq m \Rightarrow m = n$

Theorem 2(b):

The order of every element of a finite group is finite.

Proof:

Let **G** be a finite group and let $a \in$ **G**. We consider all positive integral powers of a, i.e.,

$$a\ ,\ a^2\ ,\ a^3,\ a^4,\ \ldots\ldots$$

Every one of these powers must be an element of G. But G is of finite order. Hence these elements can not all be different. We may therefore suppose that

$$a^s = a^r,\ s > r\ .$$

Now, $a^s = a^r$

$\Rightarrow$ $a^s.a^{-r} = a^r.a^{-r}$.

$\Rightarrow$ $a^{s-r} = a^o$

$\Rightarrow$ $a^{s-r} = e$

$\Rightarrow$ $a^t = e$ (putting $s - r = t$).

Since, $s > r$, t is a positive integer.

Hence there exists a positive integer t such that $a^t = e$

Now, we know that every set of positive integers has a least number. It follows that the set of all those positive integer t such that $a^t = e$ has a least member, say m. Thus there exists a least positive integer m such that $a^m = e$, showing that the order of every element of a finite group is finite.

Theorem 3(a):

The order of an element of a group is the same as that of its inverse a^{-1}.

Proof:

Let n and m be the order of a and a^{-1} respectively.

Then, $a^n = e$ and $(a^{-1})^m = e$.

Now, $a^n = e \Rightarrow (a^n)^{-1} e^{-1}$

$\Rightarrow$ $(a^{-1})^n = e$

$\Rightarrow$ $o(a^{-1}) \leq n \Rightarrow m \leq n.$

Also $\Rightarrow$ $o(a^{-1}) = m \Rightarrow (a^{-1})^m = e$

$\Rightarrow$ $o(a^m) = e \Rightarrow a^m = e$ $[\because b^{-1} = e \Rightarrow b = c]$

$\Rightarrow$ $o(a) \leq m \Rightarrow n \leq m$

Now $\Rightarrow$ $m \leq n$ and $n \leq m \Rightarrow m \leq n.$

If the order of a is infinite, then the order of a^{-1} cannot be finite. Because $o(a^{-1}) = m \Rightarrow o(a) \leq m \Rightarrow o(a)$ is finite. Therefore if the order of a is infinite, then the order of a^{-1} must also be infinite.

Theorem 3(b):

The order of any integral power of an element a can not exceed the order of a.

Proof:

Let a^k be any integral power of a. Let $o(a) = n$.

Now, $o(a) = n \Rightarrow a^n = e$ (identity element)

$\Rightarrow$ $(a^n)^k = e^k$

$\Rightarrow$ $a^{nk} = e \Rightarrow (a^k)^n = e$

$\Rightarrow$ $0(a^k) \leq n.$

Theorem 3(c):

If the element a of a group **G** *is of order n, then $a^m = e$ iff n is a divisor of m.*

Proof:

Since m must be greater than n,

Let $m = nq + r$, where $o \leq r < n$.

Now $a^m = a^{nq+r} + r = (a^n)^q . a^r = a^r$.

Since $a^n = e$. Therefore $a^r = e$.

But this is not possible for $o < r < n$, since n is the least integer for which $a^n = e$.

It follows that $r = 0$, i.e., $m = nq$.

Conversely, if $m = nq$, then $a^m = a^{nq} = (a^n)^q = e^q = e$. More generally, if m is a an integer (not necessarily positive), then also $m = nq$, where q also is an integer. In either case we say that m is a multiple of n, or that n is a divisor of m.

CYCLIC GROUP

Definition : *A group G is called cyclic if, for some $a \in G$, every element $x \in G$ is of the form a^n, where n is some integer. The element a is then called a generator of G.*

There may be more than one generators of a cyclic group. If **G** is a cyclic group generated by a, then we shall write **G** = {a} or **G** = (a). The elements of **G** will be of the form..., a^{-3} , a^{-2} , a^{-1}, a^0 (= e), a , a^2, a^3,... of course they are not necessarily all distinct.

Examples:

(i) The multiplicative group $\{1, \omega, \omega^2\}$ is cyclic. The generators are ω and ω^2.

(ii) The multiplicative group G= (1, –l, i, –i} is cyclic. We can write :
$G = \{i, i^2, i^3, i^4\}$. The generators are i and –i.

(iii) The multiplicative group of n n^{th} roots of unity is cyclic, a generator being $e^{2\pi/n}$.

Properties of Cyclic Groups

Theorem 1:

The generator of a cyclic group of order n are all the elements a^p, p being prime to n and $o < p < n$.

Proof:

We know that

$$(a^p)^n = (a^n)^p = e \text{ , therefore the order of } a^p \text{ is n.}$$

Also $$(a^p)^s = a^{ps} \neq e \text{ if } o < s < n \text{ ,}$$

because n does not divide p, nor does it divides s, therefore it does not divide ps.

Now, let $$ps = nq + r, \ o < r < n$$

and
$$(a^p)^s = a^{ps} = a^{nq+r}$$
$$= (a^n)^q \ . \ a^r = e^q \ . \ a^r$$
$$= e \ . \ a^r = a^r \neq 0 \qquad \text{(since } o < r < n\text{).}$$

Thus, a^p is a generator of the group.

Theorem 2(a):

Every cyclic group is abelian.

Proof:

Let a be a generator of a cyclic group G and let a^r, $a^s \in$ **G** for any r, s $\in$ **I** then

$a^r a^s$. $a^{r+s} = a^{s+r}$ $(\because r + s = s + r$ for r. s $\in$ **I**)

$= a^s . a^r$

Thus the operation is commutative and hence the cyclic group G is abelian.

Note : For the addition composition the above proof could have been written as

$a^r a^s = ra + sa = sa + ra$ (addition of integers is commutative).

$a^s + a^r$.

Thus the operation + is commutative in **G**.

Theorem 2(b):

The order of a cyclic group is same as the order of its generator.

Proof:

Let the order of a generator a of a cyclic group be n, then

$$a^n = e \text{ while } a^s \neq e \text{ for } o < s < n$$

When $s > n, s = nq + r$, $o \leq r < n$ (say), we observe that

$$a^s = a^{nq+r} = (a^n)^q . a^r = e^q . a^r$$

$$= e . a^r = a^r.$$

Thus there are exactly n elements in the group by a^r, where $o \leq r < n$. Therefor there are n and only n distinct elements in the cyclic group, i.e., the order of the group is n.

SUBGROUPS

Let G be a group and H any subset of G. Let a, b be any two elements of H. Now a, b being members of G the product of a b surely belongs to G, but it may or may not belongs to H. If, however, a b belongs to H, we say that H is *stable* for the composition in **G** and that the composition in G has induced the composition in H. If H is itself a group for the *induced* composition, then we say that **H** is a *subgroup* of **G**.

Definition: *A non-empty subset H of a group G is said to be a subgroup of G if the composition in G induces a composition in H and if H is a group for the induced composition.*

The two sub-groups (i) consisting of the identity element alone, and (ii) the group G itself are always present in a group G.

These are, however, *trivial* subgroups. A sub-group other than these two is known as *proper sub-group.*

A *complex* is any subset of a group, whether it is a sub-group or not.

It is easy to prove that

(i) the identity of a sub-group is the same as that of the group.

(ii) the inverse of any element of a sub-group is the same as the inverse of the element regarded as a member of the group,

(iii) the order of any element of a sub-group is the same as that of the element regarded as a member of the group.

Examples:

(i) The additive group of integers is a sub-group of the additive group of rational numbers.

(ii) The multiplicative group of positive rational numbers is a sub-group of the multiplicative group of non-zero real numbers.

(iii) The multiplicative group $\{1, -1\}$ is a sub-group of the multiplicative group $\{1, -1, i, -i\}$.

Necessary and Sufficient Condition

The necessary and sufficient conditions for a subset of a group to be a sub-group are stated in the following two theorems.

Theorem 1:

A subset H of a group G is a sub-group iff

$$(i)\ (a \in H.\ b \in H) \Rightarrow a \circ b \in H.$$

and

$$(ii)\ a \in H \Rightarrow a^{-1} \in H.$$

Proof:

Suppose H is a sub-group of G then H must be closed with respect to composition o in G, i.e., $a \in H, b \in H \Rightarrow a \circ b \in H$.

Let $a \in H$ and a^{-1} be the inverse of a in G. Then the inverse of a in H is also a^{-1}. As H itself is a group, each element of H will possess inverse in it, i.e.,

$$a \in H \Rightarrow a^{-1} \in H.$$

Thus the condition is necessary. Now let us examine the sufficiency of the condition.

(i) **Closure axiom :** $a \in H$. $b \in H \Rightarrow a \text{ o } b \in H$. Hence **closure axiom** is satisfied with respect to the operation o.

(ii) **Associativity :** Since the elements of H are also the elements of G, the composition is associative in H also.

(iii) **Existence of identity :** The identity of the subgroup is the same as the identity of the group because,

$a \in H \Rightarrow a^{-1} \in H$ [given condition (ii)]

and $a \text{ o } a^{-1} \in H$ [given condition (i)]

i.e., $e \in H$

$\therefore$ The identity e is an element of H.

(iv) **Existence of inverse :** Since $a \in H \Rightarrow a^{-1} \in H, \forall a \in H$.

Therefore each element of H possesses inverse.

Thus H itself is a group for the composition in G. Hence H is a sub-group.

Theorem 2:

***A** necessary **and** sufficient condition for a non-empty subset H of a **group** G to **sub-group** is that $a \in H$, $b \in H \Rightarrow a \text{ o } b^{-1} \in H$ where b–1 is the **inverse** of **b** in G.*

Proof:

The condition is necessary. Suppose H is a sub group of G and let

$$a \in H, b \in H.$$

Now each element of H must possess inverse because H itself is a group.

$$b \in H \Rightarrow b^{-1} \in H.$$

Also H is closed under the composition o in G. Therefore

$$a \in H, b^{-1} \in H \Rightarrow a \text{ o } b^{-1} \in H.$$

The condition is sufficient. It is given that $a \in H$, $b \in H \Rightarrow a \text{ o } b^{-1} \in H$ then we have to prove that H is a sub group.

(i) **Closure property :** Let a, $b \in H$ then $b \in H \Rightarrow b^{-1} \in H$ (as shown above)

Therefore by the given condition

$a \in H$. $b^{-1} \in H \Rightarrow a \text{ o } (b^{-1})^{-1} \in H$.

$\Rightarrow a \text{ o } b \in H$.

Thus H is a closed with respect to the composition o in G.

(ii) **Associative property :** Since the elements of H are also the elements of G, the composition is associative in H.

(iii) **Existence of Identity :** Since

$a \in H, a^{-1} \in H \in a \text{ o } a^{-1} \in H$

$\Rightarrow e \in H.$

Thus the identity element belongs to H.

(iv) **Existence of Inverse :** Let $a \in H$ then

$e \in H.\ a \in H \Rightarrow e \text{ o } a^{-1} \in H$

$\Rightarrow a^{-1} \in H.$

Thus each element of H possesses inverse.

Hence H itself is a group for the composition o in group G.

Properties of Subgroups

Theorem 1:

The union of two sub-groups is a sub-group if and only if one is contained in the other.

Proof:

Let H_1 and H_2 be two sub-groups of a group G.

(i) Let $H_1 \subset H_2$ or $H_2 \subset H_1$.

Then $H_1 \cup H_2 = H_2$ or H_1

But H_1, H_2 are sub-groups so that $H_1 \cup H_1$ is also a sub-group.

(ii) Next suppose $H_1 \cup H_2$ is a sub-group.

To prove that $H_1 \subset H_2$ or $H_2 \subset H_1$, Assume if possible that

$H_1 \not\subset H_2, H_2 \not\subset H_1$

Now, $H_1 \not\subset H_2 \Rightarrow \exists\ s \in H_1$ and $s \notin H_2$ ...(i)

and $H_2 \subset H_1 \Rightarrow \exists\ t \in H_2$ and $t \notin$ ...(ii)

From (i) and (ii), it follows that

$s \in H_1 \cup H_2, t \in H_1 \cup H_2$

since $H_1 \cup H_2$ is a sub-group, we see that

$st = k$ (say) is also an element of $H_1 \cup H_2$.

But $st = k \in H_1 \cup H_2$

$\Rightarrow$ $st = k \in H_1$ or H_2

Suppose $st = k \in H_1$

then $t = s^{-1} k \in H_1$ [$\because$ H_1 is a sub-group, $s^{-1} \in H_1$]

This contradicts (ii) Hence either $H_1 \subset H_2$ or $H_2 \subset H_1$

Theorem 2(a):

The intersection of two sub-groups of a group G is a sub-group of G

Proof:

Let H_1 and H_2 be any two sub-groups of G.

Then, $H_1 \cap H_2 \neq \phi$ because at least the identity element e is common in both H_1 and H_2.

Now to prove that $H_1 \cap H_2$ is a subgroup of G, it is sufficient to show that

$a \in H_1 \cap H_2$, $b \in H_1 \cap H_2 \Rightarrow aob^{-1} \in H_1 \cap H_2$, o being composition in G.

Since $a \in H_1 \cap H_2 \Rightarrow a \in H_1$ and $a \in H_2$ and $b \in H_1 \cap H_2 \Rightarrow b \in H_1$ and $b \in H_2$ and H_1, H_2 are sub-groups of G, we see that

$$a \in H_1,\ b \in H_1 \Rightarrow a \text{ o } b^{-1} \in H_1$$

and similarly $a \in H_2$, $b \in H_2 \Rightarrow a \text{ o } b^{-1} \in H_2$

Thus, $a \text{ o } b^{-1} \in H_1$, $a \text{ o } b^{-1} \in H_2$

$$\Rightarrow a \text{ o } b^{-1} \in H_1 \cap H_2$$

Hence $a \in H_1 \cap H_2$, $b \in H_1 \cap H_2 \Rightarrow a \text{ o } b^{-1} H_i \cap H_2$, which establishes that $H_1 \cap H_2$ is a sub-groups of G.

Theorem 2(b):

The union of two sub-groups is not necessarily a sub group.

Proof:

For example, let G he the additive group of integers, and let

$$H_1 = \{0, \pm 2, \pm 4, \pm 6, ...\}$$

$$H_z = \{0, \pm 3, \pm 6, \pm 9, ...\}$$

Then H_i, H_2 are sub-groups of G, but

$$H_1 \cup H_2 = \{0, \pm 2, \pm 3, \pm 4, \pm 6, ...\}.$$

which is not a group. It is evident that the closure property is not satisfied. For, 2 + 3 = 5, which does not belong to $H_1 \cup H_2$.

The set $H_1 \cap H_2 = \{0, \pm 6, \pm 12, ...\}$

which is certainly a group.

Sub-groups of Cyclic Groups

Theorem 1:

Every sub-group of on infinite cyclic group is infinite.

Proof:

Let G = {a} be infinite cyclic group. Let H be a sub-group of G. Then by the preceding theorem, H = $\{a^m\}$ where m is the least positive integer such that

$a^m \in H$. Now suppose, if possible, that H is finite.

This implies that $(a^m)^s = \in$ for some $s > 0$.

It follows that a is of finite order and this in turn implies that G is finite, contrary to the hypothesis. Hence H must be infinite cyclic sub-group of G

Theorem 2:

Every sub-group of a cyclic group is cyclic.

Proof:

Let G = (a) be a cyclic group generated by a. Let H be a sub-group of G. Now every element of G, hence also of H, has the form a^s, s being an integer. Let m be the smallest possible integer such that $a^m \in H$. We claim that H = $(a^m\}$. For this it is sufficient to show that $a^s \in H$, then s = mh for then $a^s = (a^m)^h$. Now, if m does not divides, then there exist integers q and r such that

$$s = mq + r, o \leq r < m.$$

Then $a^s = am^{q+r} = a^{mq}\, a^r$

or $a^r = a^s. (a^{mq})^{-1}$ ...(i)

Since $a^m \in H$, it follows that $a^{mq} \in H$ and hence its inverse $(a^{mq})^{-1} \in H$.

But $a^s \in H$ by supposition. Then from (i) it follows that $a^r \in H$ contrary to the choice of m since m was assumed to be the least positive integer such that

$a^m \in H$. Therefore $r = 0$

and so $s = mq$

But then $a^s = (a^m)^q$

Thus every element a^s of H is of the form $(a^m)^q$. Hence H = $\{a^m\}$.

ALGEBRA OF COMPLEXES OF A GROUP

Let us consider the set of all complexes of a group G, which is nothing but power set of G. Let it be denoted by P (G). Now, we define three binary compositions in P (G). The two compositions namely union and intersection of sets are familar ones. Here we define the multiplication of complexes.

Multiplication of Complexes

Let H and K be two complexes of a group G whose composition has been denoted multiplicatively, then the product of H and K denoted by HK is defined as

$$HK = \{hk : h \in H, \text{ and } k \in K\}.$$

In other words HK is the set of all possible products of elements of H with those of K.

It is evident that $h\,k \in H\,K$

$\Rightarrow h \in H, k \in K$

$\Rightarrow h, k \in G$ $\quad (\because H \subset G, K \subset G)$

$\Rightarrow h\,k \in G$ $\quad$ (Closure law in (i))

$\therefore HK \subset G$

Thus the product of two complexes is also a complex of the group.

Multiplication of Complexes is Associative

Let H, K and L be three complexes of a group G whose composition is denoted multiplicatively. Then

$$HK = \{hk : h \in H, k \in K\}$$

$$\therefore \quad (HK)\,L = \{(h\,k)\,1 : h \in H, k \in K, 1 \in L\}$$

$$= \{h\,(k\,l) : h \in H, k \in K, 1 \in L)$$

$[\because (h\,k)\,1 = h\,(k\,1)$, multiplication in G being associative]

Also $\quad H\,(K\,L) = \{h\,(k\,I) : h \in H, k \in K, 1 \in L\}$

$\therefore \quad (IIK)\,L = H\,(KL).$

Inverse of a Complex in a Group

Let H be any complex of G and let us define

$$H^{-1} = \{h^{-1} : h \in H)$$

then H^{-1} is the complex of G consisting of the inverse of the elements of H. This H^{-1} is called the inverse of complex H.

Theorem 1:

If H is any subgroup of G then $H^{-1} = H$. Also show that the converse is not true.

Proof:

Let $h^{-1} \in H^{-1}$ then $h \in H$.

Since H is a sub-group of G.

$h \in H \Rightarrow h^{-1} \in H$ (Inverse axiom)

Therefore, $h^{-1} \in H^{-1} \Rightarrow h^{-1} \in H$

$\therefore$ $H^{-1} \subseteq H$. ...(i)

Again $h \in H \Rightarrow h^{-1} \in H$

$\Rightarrow$ $h^{-1} \in H^{-1}$

$\Rightarrow$ $h \in H^{-1}$

$\therefore$ $H \subseteq H^{-1}$ ...(ii)

Hence from (i) and (ii), $H^{-1} = H$.

Theorem 2:

If H and K are any two complexes of a group G, then $(HK)^{-1} = K^{-1} H^{-1}$ (reversal law).

Proof:

Let $x \in (HK)^{-1}$ then

$x \in (HK)^{-1} \Rightarrow x = (hk)^{-1}$ where $h \in H, k \in K$

$\Rightarrow x = k^{-1} h^{-1}$

$\Rightarrow x \in K^{-1} H^{-1}$

[Since $k^{-1} \in K^{-1}$ and $h^{-1} \in H^{-1}$ $\therefore$ $k^{-1} h^{-1} \in K^{-1} H^{-1}$]

$\therefore$ $(HK)^{-1} \subseteq K^{-1} H^{-1}$ ...(i)

Again, let $y \in K^{-1} H^{-1}$ then

$y \in K^{-1} H^{-1} \Rightarrow y = k^{-1} h^{-1}, k \in K, h \in H$

$\Rightarrow y = (hk)^{-1}$ [$\because$ $(hk)^{-1} = k^{-1} h^{-1}$]

$\Rightarrow y \in (HK)^{-1}$

[$\because$ $(HK)^{-1} = \{(hk)^{-1} : h \in H, k \in K,\}$] ...(ii)

$\therefore$ $K^{-1} H^{-1} \subseteq (HK)^{-1}$

From (i) and (ii)

$(HK)^{-1} = K^{-1} H^{-1}$.

COSETS

Definition : *If G is a group, if is a sub-group and a any element in G, then the set*

$$\{h\ a : h \in H\}$$

is called the right coset generated by a and H and is denoted by Ha.

Similarly, the set

$$\{a\ h : h \in H\}$$

is the left coset denote by aH.

Since e H = H e = H, we see that H itself is a right as well as a left coset. Moreover, since $e \in H$, it is evident that $a \in a\,H$.

If the group operation is 'addition', we define the right coset of H in G by

$$H + a = \{h + a : h \in H\}$$

Similarly left coset in an additive group shall be written as

$$a + H$$

It must be noticed that cosets are not necessarily sub-groups of G, They are only special types of complexes which are sometimes called *residue classes modulo sub-group.*

In general $a\,H \neq H\,a$. In the case of an abelian group each right coset coincides with the corresponding left coset.

Properties of Cosets

Theorem 1:

If Ha = Hb, where a ,b $\in$ *G, then* $ab^{-1} \in H$*, and conversely If aH = bH, then* $a^{-1} b \in H$.

Proof:

If Ha = Hb, then a belonging to Ha, is equal to some element $h_i b$ in Hb, i.e., $a = h_i b$, or $ab^{-1} = h_i \in H$

Conversely, if $ab^{-1} \in H$, then $ab^{-1} = h_i$ i.e., $a = h_i b$.

Therefore, $Ha - H (h_i b) = (H h_i)b = IIb$.

Similarly we can slow that aH = bH iff $a^{-1}b \in H$.

Theorem 2(a):

If $h \in H$*, then the right (or left) coset Hh (or hH) of H is identical with H, and conversely.*

Proof:

Let h' be an arbitrary element of H so that $hh' \in h\,H$. Again, since H is a sub-group, we have

$$h \in H, h' \in H \Rightarrow hh' \in H$$

Thus every element of hH is also an element of H.

Hence $hH \subset H$ ...(i)

Again $h' = (hh^{-1})\, h' = h\,(h^{-1}\, h') \in h\,H$

$[\because h^{-1} \in H, h' \in H \Rightarrow h^{-1}\, h' \in H]$

This shows that every element of H is also an element of hH. Hence

$H \subset hH$ ...(ii)

From (i) and (ii) it follows that

$$hH = H$$

Similarly, we can show that

$$Hh \Rightarrow H$$

Conversely, $Hh = H \Rightarrow \in H \in H \Rightarrow h \in H$ and

similarly $hH \Rightarrow H \Rightarrow h \in H$

Theorem 2(b):

Any two right (or left) cosets of H are either disjoint or identical.

Proof:

Let H be a sub-group of a group G and let aH and bH be two left cosets. Suppose these cosets are not disjoint. Then they possess an element, say c, in common. Then c may be written as c = a h, and also as c = a h', where h and h' are in H.

Therefore, $a\,h = bh'$

or $a = bh'\, h^{-1}$

Since H is a sub-group, $h'h^{-1} \in H$.

Let $h'h^{-1} = h''$

Then $a = bh''$

Hence $aH = (bh'')H$

$= b\,(h''\, H) = bH$ [Theorem 1]

Therefore the two left cosets are identical if they are not disjoint. Thus either $a\,H \cap bH = \phi$ or $a\,H = bH$.

A similar result can be shown to hold for right cosets.

Theorem 3:

If H is finite the number of elements in a right (or left) coset of H is equal to order of H.

Proof:

The mapping $f : H \to Ha$, defined by $f(h_i) = h_i a$, is one-one onto.

It is one-one since

$$f(h_i) = f(h_j) \Rightarrow h_i a = h_j a$$
$$\Rightarrow h_i = h_j$$

from the right cancellation law; it is onto, since an element has belonging to Ha is the f-image of h belonging to H.

It follows that the number of elements in a right coset of H is the same as that in H.

Similarly the number of elements in a left coset of H is the same as that in H.

Coset Decomposition

Let H be a sub-group of G. We know that no right coset of H in G is empty and any two right cosets of H in G are either disjoint or identical.

The union of all right cosets of H in G is equal to G. Hence the set of all right cosets of H in G gives a partition of G.

This partition is called *right coset decomposition* of G. The procedure to obtain distinct members of this partition is given below :

H itself is a right coset. Now suppose $a \in G$ and $a \notin H$ then Ha will be another distinct right coset. Again let b be another such element that $b \in G$ and $b \notin H$ and also $b \notin Ha$, then Hb will be another distinct right coset. Proceeding in this way all distinct right cosets of H in G will be obtained.

Thus, $G = H \cup Ha \cup Hb \cup Hc$ .., where a, b, c are elements of G so chosen that all right cosets are distinct.

In the same way left coset decomposition of G can be obtained.

Relation of Congruence Modulo a Subgroup H is a Group G

Let H be a sub-group of a group G. If the element a of G belongs to the right coset Hb, i.e., if $a \in Hb$, i.e., if $ab^{-1} \in H$ then it is said that a is congruent to b modulo H.

Definition : *Let H be a sub-group of a group G. For a, $b \in G$ we say that a is congruent to b mod H if and only if $ab^{-1} \in H$.*

Symbolically, it can be expressed as $a \equiv b \pmod{H}$ iff $ab^{-1} \in H$.

Theorem:

The relation of congruence in a group G defined by $a \equiv b \pmod{H}$ iff $ab^{-1} \in H$. is an equivalence relation.

Proof:

(i) *Reflexivity* : Let $a \in G$ than $aa^{-1} = e \in H$ because H is a sub-group of G.

Hence $a \equiv a \pmod{H}\ \forall\ a \in G$.

The relation is reflexive.

(ii) Symmetry : $a \circ b \pmod{H}$

$\Rightarrow\ ab^{-1} \in H$

$\Rightarrow\ (ab^{-1})^{-1} \in H$ [H is a sub-group of G]

$\Rightarrow\ ba^{-1} \in H$

$\Rightarrow\ b \equiv a \pmod{H}$

Hence the relation is symmetric.

(iii) Transitivity :

$a \equiv b \pmod{H}$ and $b \equiv c \pmod{H}$

$\Rightarrow\ ab^{-1} \in H$ and $bc^{-1} \in H$

$\Rightarrow\ (ab^{-1})(bc^{-1}) \in H$ (Closure property)

$\Rightarrow\ a(b^{-1})\ c^{-1} \in H$

$\Rightarrow\ ac^{-1} \in H$ [$\because bb^{-1} = c$]

$\Rightarrow\ a \equiv c \pmod{H}$.

Hence the relation is transitive.

Thus the relation congruence mod H is an equivalence relation in G.

LAGRANGE'S THEOREM

The order of a subgroup of a finite group is a dvisor of the order of the group.

Proof:

Let H be any sub-group of order m of a finite group G of order n. Let us consider the left coset decomposition of G relative to H.

We will first show that each coset aH consists of m different elements.

Let $H = \{h_1, h_2, \ldots\ h_m\}$

Then $ah_1, ah_2, \ldots, ah_m$, are the m mebers of aH all distinct.

For, we have $ah_i = ah_j \Rightarrow h_i = h_j$

by cancellation law in G

Since G is a finite group, the number of distinct left cosets will also be finite, say k. Hence the total number of elements of all cosets is k m which is equal to the total number of elements of G. Hence

$$n = mk.$$

This shows that m, the order of H, is a divisor of n, the order of the group G.

We also that the index k is also a divisor of the order of the group.

Corollary 1:

Every finite group of composite order possesses proper subgroups.

Proof:

Let G be a finite group of order $n = pq$, where $p \neq 1$, $q \neq 1$,

If G be cyclic, and a be its generator, the group generated by a^p (or a^q) is a sub-group of G of order q (or p).

The G be not cyclic, and $a \neq e \in G$, the order of a is a proper divisor of n.

Therefore the cyclic group generated by a is a proper sub-group of G.

Corollary 2:

Fermat's Theorem : If p is a prime number which does not divide the integer a, then $a^{p-1} \equiv 1 \pmod{p}$.

Proof:

The multiplicative groups of non-zero residue classes modulo p is

$$G = (\{1, 2, 3, \ldots\ldots, p-1\} \times p),$$

which is of order $p - 1$, with the identity 1.

Since a is not divisible by p, it is congruent to one of the numbers 1, 2, 3, ..., $p - 1$, modulo p.

It follows from Cor. 1, that $a^{p-1} \equiv 1 \pmod{p}$

Corollary 3(a):

If G is of finite order n, then the order of any $a \in G$ *divides the order of G and in particular* $a^n = e$.

Proof :

Let a be of order m so that m is the least positive integer such that

$$a^m = e.$$

Then it is easy to verify that the elements

$$a, a^2, a^3, \ldots\ldots a^{m-1}, a^m = e$$

of G are all distinct and form a sub-group.

Since this sub-group is of order m, it follows that m, the order of a, is a divisor of the order of the group.

We may write n = mk,

where k is a positive integer.

Then $a^n = a^{mk} = (a^m)^k = e$.

Corollary 3(b):

A finite group of prime order has no proper sub-groups.

Proof:

Let the order of the group G be a prime number p.

Since p is a prime, its only divisors are 1 and p.

Therefore the only sub-groups of G are {e} and G, i.e. the group G has no proper sub-groups.

Corollary 3:

Every group of prime order is cyclic.

Proof:

Let G be a group of prime order p, and let $a \neq e \in G$. Since the order of a is a divisor of p, it is either 1 or p.

But $o(a) \neq 1$, since $a \neq e$.

Therefore, o (a) = p, and the cyclic subgroup of G generated by a is also of order p.

It follows that G is identical with the cyclic sub-group generated by a, i.e., G is cyclic.

QUOTIENT GROUPS

Definition : *If G is a group and N is a normal sub-group of G, then the set G|N of all cosets of N in G is a group with respect to multiplication of cosets. It is called the quotient group or factor group of G by N.*

The identity element of the quotient group G|N is N.

Theorem:

The set of all cosets of a normal sub-group is a group with respect to multiplication of complexes as the composition.

Proof:

Let N be a normal sub-group of a group G. Since N is normal in G, therefore each right coset-with be equal to the corresponding left coset.

Thus there is no distinction between right and left cosets and we shall call them simply as cosets. Let G|N be the collection of all cosets of N in G, i.e., let

$$G|N = \{Na : a \in G\}.$$

Closure property : Let a, b $\in$ G.

Then $(Na)(Nb) = N(aN)b$

$= N(Na)b$ [$\because$ N is normal]

$= NNab$

$= Nab$ [$\because$ NN = N]

Since ab $\in$ G, therefore N a b is also a coset of N in G. So Nab $\in$ G|N.

Thus G|N is closed with respect to coset multiplication.

Associativity : Let a, b, c $\in$ G. Then Na Nb, Nc $\in$ G|N.

We have $Na\,[(Nb)\,(Nc)] = Na\,(Nbc)$

$= Na(bc) = N(ab)c$ [$\because$ a(bc) = (ab)c]

$= (Nab)Nc$

$= [(Na)\,(Nb)]\,Nc.$

Thus the product in G|N salsifies the associative law.

Existence of Identity : We have N = Ne $\in$ G|N. Also if N a is any element of G|N, then

$$N\,(Na) = (Ne)\,(Na) = Nea = Na$$

and similarly $(Na)\,N = (Na)\,(Ne) = Nae = Na$

Therefore the coset N is the identity element.

Existence of Inverse : Let Na $\in$ G|N, then $Na^{-1} \in$ G|N.

We have $(Na)\,(Na^{-1}) = Naa^{-1} = Ne = N.$

and $(Na^{-1})\,(Na) = Na^{-1}\,a = Ne = N.$

The coset Na^{-1} is the inverse of Na. Thus each element of G|N possesses inverse.

Hence G|N is a group with respect to product of cosets.

NORMAL SUB-GROUPS

Definition : *A sub-group N of a group G is said to be a normal sub-group of G if for every* $x \in G$ *and for every* $n \in N$, $x\,n\,x^{-1} \in N$.

From this definition we can immediately conclude that N is a normal subgroup of G if and only if

$$xNx^{-1} \subset N \ \forall \ x \in G.$$

Theorem 1:

A sub-group N of a group G is a normal sub-group of G if and only if each left coset of N in G is a right coset of N in G.

Proof:

Let N be a normal sub-group of G.

Then $xNx^{-1} = N \forall \ x \in G$

$\Rightarrow$ $(xNx^{-1})x = Nx$ for all $x \in G$

$\Rightarrow$ $xN = Nx$ for all $x \in G$

$\Rightarrow$ each left coset xN is the right coset NX.

Theorem 2:

A sub-group N of a group G is normal if and only if

$$xNx^{-1} = N \ \forall \ x \in G$$

Proof:

Let $xNx^{-1} = N \forall \ x \in G$

Then $xNx^{-1} \subset N \forall \ x \in G$. Therefore N is a normal sub-group of G.

Conversely, Let N be a normal sub-group of G.

Then $xNx^{-1} \subset N \forall \ x \in G$ ...(i)

Also $x \in G \Rightarrow x^{-1} \in G$. Therefore we have $x^{-1}N(x^{-1})^{-1} \subset N \ \forall \ x \in G$

$\Rightarrow$ $x^{-1}Nx \subset N \forall \ x \in G$

$\Rightarrow$ $x(x^{-1}Nx)x^{-1} \subset xNx^{-1} \ \forall \ x \in G$

$\Rightarrow$ $N \subset x \, N \, x^{-1}$ for all $x \in G$ ...(ii)

From (i) and (ii), we conclude that

$$xNx^{-1} = N \text{ for all } x \in G.$$

Conversely, let each left coset of N in G be a right coset of N in G. It means that if x is any element of G, then the left coset x N is also a right coset. Now $e \in N$ and therefore $xe = x \in xN$. So x must also belong to that right coset which is equal to the left coset x N. But x is an element of the right coset Nx and two right cosets are either disjoint or identical. Therefore Nx is the unique right coset which is equal to the left coset x N. Therefore, we have

$xN = Nx \ \forall \ x \in G$

$\Rightarrow$ $xNx^{-1} = Nxx^{-1} \ \forall \ x \in G$

$\Rightarrow$ $xNx^{-1} = N \ \forall \ x \in G$

$\Rightarrow$ N is a normal sub-group of G.

Theorem 3(a):

The intersection of two normal sub-groups of a group is a normal sub-group.

Proof:

Let H and K be two normal sub-groups of a group G. Since H and K are sub-groups of G, therefore $H \cap K$ is also a sub-group of G. Now to prove that $H \cap K$ is a normal sub-group of G.

Let x be any element of G and n be any element of $H \cap K$.

We have $n \in H \cap K \Rightarrow n \in H, n \in K$.

Since H is a normal sub-group of G, therefore

$$x \in G, n \in H \Rightarrow xnx^{-1} \in H$$

Similarly, $xn x^{-1} \in K$.

Now, $xnx^{-1} \in H, xnx^{-1} \in K$

$\Rightarrow$ $x n x^{-1} \in H \cap K$.

Thus, we have $x \in G.\ n \in G\ n \in H \cap K$

$\Rightarrow$ $xnx^{-1} \in H \cap K$

$\Rightarrow$ $H \cap K$ is a normal sub-group of G.

Theorem 3(b):

A sub-group N of a group G is a normal sub-group of G if and only if the product of two right cosets of N in G is again a right coset of N in G.

Proof:

Let N be a normal sub-group of a group G. Let a, b be any two elements of G. Then Na and Nb are two right cosets of N in G. We have

$$\begin{aligned}(Na)(Nb) &= N(aN)b \\ &= N(Na)b \qquad [\because N \text{ is normal} \Rightarrow Na = aN] \\ &= NNab \\ &= Nab\end{aligned}$$

Since $a \in G, b \in G \Rightarrow ab \in G$, therefore N a b is also a right coset of N in G. Thus the product of the right cosets Na and Nb is the right coset Nab.

Conversely, let N be a sub-group of G such that the product of two right cosets of N in G is again a right coset of N in G. Let x be any element of

G. Then $x^{-1} \in G$. Therefore Nx and Nx^{-1} are two right cosets of N in G. Consequently, by hypothesis $NxNx^{-1}$ is also a right coset of N in G. Since $e \in N$, therefore, $e\ x\ e\ x^{-1} = e$ is an element of the right coset $NxNx^{-1}$. But N itself is a right coset of N in G and $e \in N$. Also if two right cosets have one element common, then they must be identical. Therefore we must have

$$NxNx^{-1} = N \forall x \in N$$

$\Rightarrow$ $n_1 x\, n x^{-1} \in N \ \forall\ x \in G$ and $n_1, n \in N$

$\Rightarrow$ $n_1^{-1} (n_1\, x\, n\, x^{-1}) \in n_1^{-1} N \ \forall\ x \in G$ and $\forall$, $n_1, n \in N$

$\Rightarrow$ $xnx^{-1} \in N \ \forall\ n_1, n \in N$

$\Rightarrow$ $x\, nx^{-1} \in N\ V\ x \in G$ and $\forall\ n \in N$

[$\because$ $n_1^{-1} N = N$ as $n_1^{-1} N$ since $n_1 \in N$]

$\Rightarrow$ N is a normal sub-group of G.

Centre of a Group

Definition: *The set Z of all, those elements of a group G which commute with every element of G is called the centre of G. Symbolically*

$$Z = [z \in G : zx = xz \Rightarrow x \in G]$$

Theorem:

The centre Z of a group G is a normal sub-group of G.

Proof:

We have $Z = [z \in G: zx = xz\ \forall\ x \in G]$

First we shall prove that Z is a sub-group of G.

Let $z_1, z_2 \in Z$. Then

$z_1 x \Rightarrow xz_1$ and $z_2 x = xz_2$ for all $x \in G$.

We have $z_2 x = xz_2\ \forall\ x \in G$

$\Rightarrow$ $z_2^{-1} (z_2 x) z_2^{-1} = z_2^{-1} (xz_2)\ z_2^{-1}$

$\Rightarrow$ $x z_2^{-1} = z_2^{-1}\ \forall\ x \in G$

$\Rightarrow$ $z_2^{-1} \in Z$

Now $(z_1\ z_2^{-1})\, x = z_1\, (z_2^{-1}\ x) = z_1\, (x z_2^{-1})$

$= (z_1 x) z_2^{-1} = (xz_1)\ z_2^{-1}$

$= x(z_1 z_2^{-1})$

$\therefore$ $z_1 z_2^{-1} \in Z.$

Thus $z_1, z_2 \in Z \Rightarrow z_1\ z_2^{-1} \in Z$.

$\therefore$ Z is a sub-group of G.

Now, we shall show that Z is a normal sub-group of G. Let $x \in G$ and $z \in Z$. Then

$$xzx^{-1} = (xz)\,x^{-1} = (zx)\,x^{-1}$$
$$= z \in Z.$$

Thus, $x \in G, z \in Z \Rightarrow x\,zx^{-1} \in Z.$

$\therefore$ Z is a normal sub-group of G.

RELATION OF CONJUGACY IN A GROUP

Conjugate Element : If $a, b \in G$, then b is said to be a conjugate of a in G if there exists an element $x \in G$ such that $b = x^{-1}\,ax$

Symbolically, we shall write $a \sim b$ for this and shall refer to this relation as conjugacy.

Thus $b \sim a \Leftrightarrow x^{-1}\,a\,x$ for some $x \in G$.

Example:

In P_3, since $(2, 3) = (1\ 2\ 3)^{-1}\,(1\ 2)\,(1\ 2\ 3)$, $(2\ 3)$ is the conjugate of $(1\ 2)$ by $(1\ 2\ 3)$.

Theorem:

Conjugacy is an equivalence relation in a group.

Proof:

(i) *Reflexivity:* Let $a \in G$, then $a = e^{-1}\,a\,e$, hence $a \sim a\ \forall\ a \in G$ i.e., the relation of conjugacy is reflexive.

(ii) *Symmetry :* Let $a \sim b$, so that there exists an $x \in G$ such that

$$a = x^{-1}\,bx,\ a\ ,\ b \in G.$$

Now $a \sim b \Rightarrow a = x^{-1}bx$

$\Rightarrow\quad x\,a = x\,(x^{-1}\,bx)$

$\Rightarrow\quad xax^{-1} = (xx^{-1})\,b\,(xx^{-1})$

$\Rightarrow\quad b = xax^{-1}$

$\Rightarrow\quad b - (x^{-1})^{-1}\,ax^{-1},\ x^{-1} \in G$

$\Rightarrow\quad b \sim a$

Thus $a \sim b \Rightarrow b \sim a$

Hence relation is symmetric.

(iii) *Transitivity :* Let there exist two elements $x, y \in G$ such that $a = x^{-1}\,b\,x$, and $b = y^{-1}\,cy$ for $a, b, c \in G$.

Hence a ~ b , b ~ c

⇒ $a = x^{-1}\ bx$ and $b = y^{-1}\ cy$

⇒ $a = x^{-1}\ (y^{-1}\ cy)\ x$

= $a = (x^{-1}y^{-1})\ c\ (yx)$

⇒ $a = (yx)^{-1}\ c\ (yx)$

Where yx ∈ G, G being the group.

Therefore, a ~ b, b ~ c ⇒ a ~ c.

Hence relation is transitive.

Thus conjugacy is equivalence relation on G.

Conjugate classes : For a ∈ G, C (a) = {x : x ∈ G and a ~ x}, C (a), the equivalence class of a in G under conjugacy relation is usually called the conjugate class of a in G. It consists of the set of all distinct elements of the type y^{-1} ay as y ranges over G.

HOMOMORPHISM

By homomorphism we mean a mapping from one algebraic system to a like algebraic system which preserves structure.

Definition: *Let G and G' be any two groups with binary operation 'o' and 'o' respectively. Then a mapping f : G → G said to be a homomorphism if for all a, b ∈ G, f(aob) = f(a) o'f(b).*

A homomorphism f, which at the same time is also onto is said to be an *epimorphism.*

A homomorphism f which is also one-one is called a *monomorphism.*

A group G' is called a homomorphic image of a group G, if there exists a homomorphism f of g onto G.

A homomorphism of a group G into itself is called an *endomorphism.*

Examples:

(i) Let G be any group under binary operation o. If f (x) = x for every x ∈ G] then f : G → G' is a homomorphism because

$$f(xy) = xy = f(x)\ f(y).$$

(ii) Let G be the group of integers under addition, let G' be the group of integers under addition modulo n. If f : G → G' be defined by f = remainder of x on division by n, then this is a homomorphism.

(iii) Let G be any group under addition. If f(x) = e ∀ x ∈ G then the mapping f : G → G', is a homomorphism because for all x, y ∈ G, f(x + y) = e and f(x) + f(y) = e + e = e, so that

$$f(x + y) = f(x) + f(y)$$

(iv) Let G be the group of integers under addition and let G' = G. If for all $x \in G$, $f(x) = 2x$ then f is a homomorphism because

$$f(x + y) = 2(x + y) = 2x + 2y = f(x) + f(y)$$

Kernel of Homomorphism

Definition : *If f is a homomorphism of a group G into a group G', then the set K of all those elements of G which are mapped by f onto the identity e' of G' is called the kernel of the homomorphism f.*

Theorem:

Let G and G' be any two groups and let e and e' be their respective identities. If f is a homomorphism of G into G' then

(1) $f(e) = e'$,

(2) $f(x^{-1}) = [f(x)]^{-1}$ *for all* $[x \in G.]$

(3) K is a normal subgroup of G.

Proof:

(1) We know that for $x \in G$, $f(x) \in G'$,

$\therefore$ $f(x).e' = f(x) = f(xe) = f(x).f(e)$

and therefore by using left cancellation law, we have

$e' = f(e)$ or $f(e) = e'$

(2) Since for any $x \in G$, $xx^{-1} = e$, we get

$$f(x).f(x^{-1}) = f(xx^{-1}) = f(e) = e'$$

Similarly $x^{-1}x = e$, gives $f(x^{-1}).f(x) = e'$

Hence by definition of $[f(x)^{-1}]$ in G' we obtain the result

$$f(x^{-1}) = [f(x)]^{-1}$$

(3) Since $f(e) = e'$, $e \in K$. This shows that $K \neq \phi$

Now let $a, b \in K$, $x \in G$, $a \in K$, $b \in K$.

$\Rightarrow$ $f(a) = e'$ $f(b) = e'$

$\Rightarrow$ $f(a) = e'$, $f(b^{-1}) = [f(b)]^{-1} = e'$

$\Rightarrow$ $f(ab^{-1}) = f(a)\,[f(b)F' = e'\,.\,e' = e'$

$\Rightarrow$ ab^{-1} K.

This establish that K is a sub-group of G.

Now, to show that it is also normal we prove the following.

$$f(x^{-1}ax) = f(x^{-1})f(a)f(x)$$
$$= [f(x)]^{-1} f(a)f(x)$$
$$= [f(x)]^{-1} e'f(x) = [f(x)]^{-1} f(x) = e'$$

$\therefore$ $x^{-1}\, a\, x \in K$.

Hence the result.

CAYLEY'S THEOREM

Every finite group is isomorphic to a permutation group.

Proof:

Let G be a finite group of order n. If $a \in G$, then $\forall\, x \in G$, $ax \in G$. Now consider a function from G into G defined by

$$f_a(x) = ax \;\forall\, x \in G$$

For, $x, y \in G$, $f_a(x) = f_a(y) \Rightarrow ax = ay$

$\Rightarrow$ $x = y$ (by left cancellation law)

Therefore, function f_a is one-one

The function f_a is also onto because if x is any element of G then there exists an element f_a such that

$$f_a(a^{-1}x) = a(a^{-1}x) = (aa^{-1})x = ex = x.$$

Thus f_a is one-one from G onto G. Therefore, f_a is a permutation on G. Let G' denote the set of all such one-to-one onto functions defined on G corresponding to every element of G, i.e.,

$$G' = \{f_2 : a \in G)$$

Now, we show that G' is a group with respect to the product of functions.

(i) **Closure axiom** : Let f_a, $f_b \in G'$, where $a, b \in G$, then

$$(f_a \circ f_b)\, x = f_a[f_b(x)] = f_a(bx) \Rightarrow a(bx)$$
$$= (ab)\, x = f_{ab}(x) \;\forall\, x \in G$$

$$f_a \circ f_b = f_{ab} \qquad ...(i)$$

Since $ab \in G$, therefore $f_{ab} \in G'$ and thus G' is closed under the product of functions.

(ii) **Associative axiom** : Let f_a, f_b, $f_c \in G'$ where $a, b, c \in G$

Then

$$f_a \circ (f_b \circ f_c) = f_a \circ f_{bc} \qquad \text{(From (i)]}$$
$$= f_a(bc) \qquad \text{[From (i)]}$$

$= (ab)\,c$ [by associative law in G]

$= f_{ab} \circ f_c$ [From (i)]

$= (f_a f_b) \circ f_c$ [From (i)]

$\therefore$ Product of functions is associative in G',

(iii) **Identity axiom :** If e the identity element in G, then G is the identity of G' because $\forall\ f_x \in G'$ we have $f_e \circ f_x = f_{ex} = f_x$

and $f_x \circ f_e = f_{xe} = f_x$.

(iv) **Inverse axiom :** If a^{-1} is the inverse of a in G, then f_a^{-1} is the inverse. of f_a in G' because

$f_a^{-1} \circ f_a = f_{aa}^{-1} = f_e$ and $f_a \circ f_a^{-1} = f_{aa^{-1}} = f_e$

Hence G' is a group with respect to composite of functions denoted by the symbol o.

Now consider the function g from G into G' defined by

$g(a) = f_a\ \forall\ a \in G$

g is one-one because for a, b, $\in$ G

$g(a) = g(b) \Rightarrow f_a = f_b$

$\Rightarrow f_a(x) = f_b(x)$

$\Rightarrow ax = bx \Rightarrow a = b,\ \forall\ x \in G$

g is onto because if $f_a \in G'$ then for $a \in G$,we have

$g(a) = f_a$

g preserves composition in G and G' because if a b $\in$ G then

$g(ab) = f_{ab}$ [by definition of g]

$= f_a \circ f_b$ [from (i)]

$= g(a) \circ g(b)$

$\therefore\quad G \cong G'$.

ISOMORPHISM

Definition : Let G, G' be two group with binary operations o and o' respectively. If there exists a one-one onto mapping : G $\to$ G' such that

$$f(a \circ b) = f(a) \circ' f(b), \text{ for all } a, b \in G,$$

then the group G is said to be isomorphic to the group G', and the mapping f is said to be an isomorphism. If G is isomorphic to G', we write

$$G \simeq G' \text{ or } G \cong G'.$$

In other words, a group G is isomorphic to the group G' if there exists one-one mapping of G to G' such that the *image of the product of two elements is the product of the images of the elements,* with respect to the compositions in the respective groups.

The last condition may also be stated as follows :

If $ab = c$, where $a, b, c \in G$,

and $f(a) = a'$, $f(b) = b'$, $f(c) = c'$,

then $a'b' = c'$, where $a', b', c' \in G'$

Properties of Isomorphism

Theorem 1:

The relation of isomorphism in the set of groups is an equivalence relation.

Proof:

Reflexivity : It can be easily shown that every group G is isomorphic to itself, and isomorphism being the identity mapping defined by

$$f : G \to G : f(x) = x \;\forall\; x \in G.$$

The mapping is evidently one-one onto,

Moreover, we have

$$f(xy) = xy, \text{ by definition of } f$$
$$= f(x)\, f(y).$$

The f is composition preserving as well.

Hence $G \cong G$.

Symmetry : Let a group G isomorphic to another group G' i.e.

$$G \cong G'.$$

and let f be an isomorphic mapping of G onto G'. Since f is a one-one onto mapping, it is invertible, f^{-1} exists and is also one-one onto. We shall show that f^{-1} is an isomorphic mapping of G' onto G.

Let $x', y' \in G'$

Then there exist elements $x, y \in G$ such that

$$f^{-1}(x') = x,\; f^{-1}(y') = y \qquad \text{...(i)}$$

so that $x' = f(x)$ and $y' = f(y)$ ...(ii)

Also since f is an isomorphism of G onto G', we have

$$f(xy) = f(x)\, f(y) \;\forall\; x, y \in G \qquad \text{...(iii)}$$

Now $f^{-1}(x'y') = f^{-1}[f(x)\, f(y)]$ by (ii)

$= (f^{-1} of)(xy)$ by (iii)

$= xy$ since f^{-1} of is the identity mapping

$= f^{-1}(x')\, f^{-1}(y')$ by (i)

Hence $G' \cong G$.

Transitivity : Let G, G', G" be three groups such that $G \cong G'$ and $G' \cong G''$ let f and ϕ be the respective isomorphic mappings.

Since f, ϕ are one-one onto ϕ of is also one-one onto.

Moreover $[\phi\, of](xy) = \phi\{f(xy)\}$, $x, y \in G$

$= \phi\{f(x)\, f(y)\}$, ($\because$ f is an isomorphism)

$= [(\phi\, of)(x)]\,[(\phi\, of)(y)]$.

Hence f is an isomorphic mapping of G onto G", so that

$G \cong G''$.

Thus the theorem is completely established.

Theorem 2(a):

If isomorphism exists between two groups, then the identities correspond, i.e., if $f : G \to G'$ is an isomorphism and e, e', are respectively the identities in G, G' then $f(e) = e'$.

Proof:

Let e be the identity in G, and let f (e) = e'. Then we well prove that e' is the identity in G'.

Let $x' \in G'$.

Since f is one-one onto, there is an $x \in G$ such that $f(x) = x'$ since f is an isomorphism. Thus

$$x' = f(x) = f(e)\, f(x) = e'\, x'$$

for every $x' \in G$,

if follows that e' is the identity in G'

Theorem 2(b):

If isomorphism exists between two groups, then the inverses correspond. i.e. if $f : G \to G'$ is an isomorphism and $f(a) = a'$ where $a \in G$, $a' \in G'$ then $f(a^{-1}) = a^{-1} = [f(a)]^{-1}$

Proof:

Let $f(a) = a'$ where $a \in G$, $a' \in G'$. We will prove that the inverse of a' is $f(a^{-1})$

Since $aa^{-1} = e$ the identity G, we have

$$f(a) . f(a^{-1}) = f(a. a^{-1})$$

$$= f (e) = e' \text{ the identity in } G'$$

i.e., $a'. f (a^{-1}) = e'$

if follows that $a^{-1} = f (a^{-1})$.

Theorem 2(c):

In an isomorphism the order of an element is preserved, i.e. if $f : G \rightarrow G'$ is an isomorphism, and the order of a is n, then the order of $f (a)$ is also n.

Proof:

As $f (a) = a'$ then we have

$$f(a^2) = f(a.a) = f(a). f(a) = a', a' = a'^2$$

and in general,

$$f(a^n) = a'^n$$

But $\quad f (a^n) = f (e) = e'$, (by Theorem 1)

Therefore $\quad a'^n = e'$

Also $\quad a'^m \neq e'$ for $m < n$ i.e., $o (a') = n$

It follows that the order of an element in G, if finite, is equal to the order of its image in G'.

If the order of a is infinite, we can similarly show that the order of a' can not be finite,

Isomorphism of Cylcic Groups

Theorem 1(a):

A subgroup of the infinite cyclic group is isomorphic to the additive group of integral multiples of an integer.

Proof:

Let $G = \{..., a^{-2}, a^{-1} = a^0 = e, a^1, a^2, a^3, ...\}$ and let H be a subgroup of G, given by

$$H = \{..., a^{-2m}, a^{-m}, a^0 = e, a^m, a^{2m},...\}$$

$$= \{(a^m)^n : n \in z\}.$$

Then H is isomorphic to the additive group H', given by

$$H' = \{0, \pm m, \pm 2m, \pm 3m,...\}$$

$= \{nm : n \in z)$

For, the mapping f: $H \to H'$ defined by $f(a^{nm}) = nm$ is one-one onto and if $r, s \in z$ then

$$f(a^{rm}, a^{sm}) = f(a^{(r+s)m}) = (r + s)^m$$

$$= rm + sm = f(a^m) + f(a^{sm})$$

It will be observed that H is itself an infinite cyclic group, and as such it is isomorphic to G. Thus a *subgroup of an infinite cyclic group is isomorphic to the group itself.*

Thcorem 1(b):

Cyclic groups of same order are isomorphic

Proof:

Let G and G' be two cyclic groups of order n, which are generated by a and b respectively. Then

$$G = \{a, a^2, a^3, ..., a^n = e\} \text{ and}$$

$$G' = \{b, b^2, b^3, ..., b^n = e'\}$$

The mapping $f : G \to G'$ defined by $f(a^r) = b^r$ is isomorphism. For, $f(a^r . a^s) = f(a^{r+s}) = b^{r+s} = b^r . b^s$

$$= f(a^r) . f(a^s)$$

Therefore the groups are isomorphic.

Theorem 2(a):

An infinite cyclic group is isomorphic to the additive group of integers.

Proof:

Let G be an infinite cyclic group generated by a. Then

$$G = \{..., a^{-2}, a^{-1} . a^0 = e, a^1, a^2, a^3 ...\}$$

$$= (a^r, \text{ is an integer})$$

The mapping f: $G \to Z$ defined by

$f(a^r) = r$ is an isomorphism. For it is one-one onto and further

$$f(a^r . a^s) - f(a^{r+s}) = r + s = f(a^r) + f(a^s)$$

If follows that G is isomorphic to Z.

Theorem 2(b):

A cyclic group of order n is isomorphic to the additive group of residue classes modulo n.

Proof:

Let G be the cyclic group of order n, generated by a. Then

$$G = \{a, a^2, a^3, \dots a^{n-1}, a^n = e\}$$

Let G' be the additive group of residue classes (mod n), i.e.,

$$G' = \{[1] . [2], \dots [n] = [0]\}$$

The mapping $f : G \to G'$ defined by $f(a^r) = [r]$ is an isomorphism.

For, it is one-one onto and further,

$$f(a^r . a^s) = f(a^{r+s}) = [r + s] = [r] + [s]$$
$$= f(a^r) + f(a^s)$$

It follows that G is isomorphic to G'.

SOLVED EXAMPLES

Example 1:

Show that the mapping f of the symmetric group onto the multiplicative group G' = {1, – 1} defined by

$$f(\alpha) = 1 \text{ or } -1.$$

according as a is an even or odd permutation in P_n is a homomorphism of P_n onto G'.

Solution:

We know that the product of two permutations both even or both odd is even while the product of one even and one odd permutations is odd. We shall show that

$$f(\alpha\, \beta) = f(\alpha)\, f(\beta)\ \forall\ \alpha, \beta \in P_n$$

(i) If α, β are both even, then

$$f(\alpha, \beta) = 1 = 1 . 1 = f(\alpha)\, f(\beta)$$

(ii) If α, β are both odd, then.

$$f(\alpha\, \beta) = 1 = (-1)\,(-1) = f(\alpha)\, f(\beta)$$

(iii) If α is odd and β is even then

$$f(\alpha\, \beta) = -1 = (-1)\,(1) = f(\alpha)\, f(\beta)$$

(iv) If α is even and β is odd, then

$$f(\alpha\, \beta) = -1 = (1)\,(-1) = f(\alpha)\, f(\beta)$$

Thus $\quad f(\alpha\, \beta) = f(\alpha)\, f(\beta)\ \forall\ \alpha, \beta \in P_n$

Also obviously f is onto G', Therefore f is a homomorphism of P_n onto G'.

Example 2:

Prove that every sub-group of an abelian group is normal

Solution:

Let H be a sub-group of an abelian group G. Let a be any element of G and h that of H.

Then we have

$aha^{-1} = aa^{-1}h,$ $[\because ha^{-1} = a^{-1} h$, as G is abelian$]$

$= eh,$ $[\because eh = h]$

$= h$

Therefore, $a \in G$, $h \in H \Rightarrow aha^{-1} \in H$ i.e., H is normal in G.

Example 3(a):

Show that there is one-one correspondence between the conjugates of an element of a group G, and the right cosets of the normalizer of a in G.

Solution:

Here N (a) is the normalizer of a in G.

Let $x, y \in G$ be in the some right coset of N (a) in G. also we know that if H is a subgroup of G, then $Ha = Hb \Leftrightarrow ab^{-1} \in H$.

$\therefore$ We have $xy^{-1} \in N(a)$

$\Rightarrow a(xy^{-1}) = (xy^{-1})a$, by definition of normalizer.

$\Rightarrow x^{-1}(axy^{-1})y = x^{-1}(xy^{-1}a)y$

$\Rightarrow x^{-1}ax = y^{-1}ay$

i.e., x and y give rise to the same conjugate of a.

Again let x, y belong to two different right cosets of N (a) in G. If possible, let $x^{-1}ax = y^{-1}ay$

$\Rightarrow x(x^{-1}ax)y^{-1} = x(y^{-1}ay)y^{-1}$

$\Rightarrow a(xy^{-1}) = (xy^{-1})a$

$\Rightarrow xy^{-1} \in N(a)$

$\Rightarrow$ x, y belong to the same right coset of N (a) in G.

But it is against the supposition that x, y belong to two different right cosets of N (a) in G.

$\therefore$ Two elements in different rights cosets of N (a) in G does not give rise to the same conjugate of a.

Example 3(b):

If $G = \{1, -1, i, -i\}$ which forms a group under multiplication and I = the group of all integers under addition. Prove that the mapping f from I onto G such that $f(x) = i^n \ \forall \ n \in I$ is a homomorphism.

Solution:

Since $f(x) = i^n$, $f(m) = i^m$, for all m, n, $\in$ I,

$$f(m + n) = i^{m+n} = i^m, i^n = f(m)f(n).$$

Hence f is a homomorphism.

Example 3(c):

Show that two elements are conjugate iff they can be put in the form ab and ba respectively where a, b $\in$ G.

Solution:

Let x, y $\in$ G be two conjugate elements, then for some z $\in$ G.

we have $\quad x = z^{-1} y z$

Now suppose that $z^{-1} y = a$ and $z = b$

Then $\quad x = z^{-1} yz = ab.$

And $\quad ba = z(z^{-1} y) = (zz^{-1}) y = ey = y$

Conversely, suppose that x = ab and y = ba

Then $y = ba \Rightarrow b^{-1} y = b^{-1} ba$

$\Rightarrow b^{-1} y = a$

$\therefore \quad x = ab \Rightarrow x = b^{-1} y b$

$\Rightarrow$ x and y are conjugate elements.

Example 4:

If H is a sub-group of index 2 in G, prove that H is a normal sub-group of G. Illustrate by an example,

Solution:

Since H is a sub-group of index 2 in G, so the number of distinct right (or left) cosets of H in G is 2.

Let a be any element of G, then if a $\in$ H, we have

$$aH = H = Ha \qquad ...(i)$$

If a $\notin$ H, then the right coset Ha as well as left coset aH are distinct from h. But H being a subgroup of index 2 in G, the number of distinct right (or left) cosets in right (or left) coset decomposition of G is 2. Hence the cosets H, aH and Ha are such that

$$G = H \cup aH = H \cup Ha$$

But there being no element common to H and aH or common to H and Ha,

we have $aH = Ha$ ...(ii)

Hence from (i) and (ii) we have $aH = Ha\ \forall\ a \in G$. i.e., H is a normal sub-group of G.

Let us consider the symmetric group P_3 on three symbol a, b, c.

i.e., $P_3 = \{f_1, f_2, f_3, f_4, f_5, f_6\}$

when f_1 = identity permutation; $f_2 = (a, b)$ $f_3 = (bc)$, $f_4 = (c\ a)$, $f_5 = (abc)$, $f_6 = (acb)$

Let A_3 = alternating group on a, b, c,

= set of even permutation belonging to P_3

Then $A_1 = \{f_1, f_5, f_6\}$

By Lagrange's Theorem A_3 will have only two distinct cosets in P_3. One in A_3 itself and the other in A_3f_2. Thus we see that the index of A_3 in P_3 is 2.

To prove that A_3 is a normal sub-group of P_3, we proceed as follows—

Let f be any element of P_3, so that f may be odd or even. Also suppose that g is any element of A_3, so that g is an even permutation. We shall prove that

$$fgf^{-1} \in A_3$$

i.e., $f\,g\,f^{-1}$ is an even permutation

If f is odd, then f^{-1} is also odd. Furthermore, fg is odd and so fgf^{-1} is even.

If f is even, then f^{-1} is also even. Further fg is even and so fgf^{-1} even.

Thus we see that $f \in P_3$, $g \in A_3 \Rightarrow fgf^{-1} \in A_3$.

Hence A_3 is a normal sub-group of P_3.

Example 5:

If N and M are normal subgroups then show that NM is also a normal sub-group of G.

Solution:

We can easily prove that NM is a sub-group of G.

Now let a be any element of G and nm that of NM.

Then $n \in N$ and $m \in M$ and $a(nm)a^{-1} = (ana^{-1})(ama^{-1})$

$\in NM$, $ana^{-1} \in N$, $ama^{-1} \in M$ as N and M are normal subgroups of G.

$\therefore$ NM is normal sub-group of G.

Example 6:

Suppose that N and M are two normal sub-groups of G such that $N \cap M = \{e\}$. Show that every element of N commutes with every element of M.

Solution:

Let x be any element of N and y be any element of M. Then we have to prove that $xy = yx$.

Let us consider the element $xyx^{-1}y^{-1}$.

Since N is normal, $yx^{-1}y^{-1} \in N$,

Also $x \in N$, so that

$$xyx^{-1}y^{-1} \in N.$$

Furthermore M is normal, $xyx^{-1} \in M$

Also $y^{-1} \in M$.

So that $xyx^{-1}y^{-1} \in M$. Thus we see that

$$xyx^{-1}y^{-1} \in N \text{ and } xyx^{-1}y^{-1} \in M.$$

This implies that $xyx^{-1}y^{-1} \in N \cap M$.

But $N \cap M = \{e\}$, so we have

$$xyx^{-1}y^{-1} = e, \text{ i.e., } xy = yx.$$

Example 7:

If ω be the imaginary cube root of unity show that the set $\{1, \omega, \omega^2\}$ is a cyclic group of order 3 with respect to multiplication.

Solution:

To verify the group axioms, let us prepare the multiplication table.

$\bullet$	1	ω	ω^2
1	1	ω	ω^2
ω	ω	ω^2	$\omega^3 = 1$
ω^2	ω^2	$\omega^3 = 1$	$\omega^4 = \omega$

From table we observe :

(i) Since all the elements in table belong to the set $\{1, \omega, \omega^2\}$, *closure axiom* is satisfied.

(ii) Multiplication is *associative.*

(iii) *Identity element* is 1.

(iv) The *inverse* of 1, ω, ω^2 are 1, ω^2, ω, to respectively.

Also, ω, ω^2 are the generators because each one of them is square of the other and $\omega^2 = 1$. Number of elements in the set is 3.

Hence the set in question is cyclic group of order 3 with respect to multiplication.

Example 8:

Find the proper subgroups of the multiplicative group G of the sixth roots of unity.

Solution:

From Trigonometry we know that six sixth roots of unity are

$\cos(n\pi/3) + i \sin(n\ \pi/3)$. i.e., $e^{n\pi i/3}$, Where n = 0, 1, 2, 3, 4, 5.

$\therefore\ G = \{e^{n\pi i/3}, n = 0, 1, 2, 3, 4, 5\}$

If H_1 and H_2 be its proper subgroups, then

$$H_1 = \{e^{n\pi i/3}, n = 3,6\} = \{e^{\pi i}, e^{2\pi i} = 1\}$$
$$= \{e^{\pi i}, 1\}$$

and

$$H_2 = \{e^{2\pi i/3}, e^{4\pi i/3}, 1\}$$
$$= \{e2^{n\pi i/3}, e^{4\pi i/3}, 1\}$$

Example 9:

If G is a group and H is its only sub-group of a given order, prove that H will be the normal sub-group of G.

Solution:

Given that H is only sub-group of G and the order of H is n.

It can easily be proved that if $x \in G$, then $x H x^{-1}$ is a sub-group of G.

Now, let $H = \{h_1, h_2, h_3, ..., h_n\}$,

then $xHx^{-1} = \{xh_1x^{-1}, xh_2x^{-1}, ... ,xh_nx^{-1})$

Also as $xh_mx^{-1} = xh_ix^{-1} \Rightarrow h_m = h_1$, the number of distinct elements in xhx^{-1} is n,

$\therefore$ order of xHx^{-1} = order of H = n.

But it is given that H is the only sub-group of G or order n, so

$$xHx^{-1} = H \ \forall\ x \in G.$$

H is a normal sub-group of G.

$$xHx^{-1} = H \ \forall\ x \in G.$$

∴ H is a normal sub-group of G.

Example 10:

Let G be the group of all ordered pairs (a, b) of real numbers with the binary operation denoted additively and defined by

$$(a, b) + (c, d) = (a + c, b + d).$$

Further let G' be the additive-group of all real numbers. Then the mapping

$$f : G \rightarrow G'$$

defined by $f(a, b) = a \ \forall\ (a, b) \in G$

is a homomorphism of G onto G'.

Solution:

It can be easily proved that G is a group with respect to the given binary operation. The ordered pair (0, 0) is the identity element and the ordered pair (–a, –b) is the inverse of (a, b).

Let (a, b) and (c, d) be any two elements of G.

Then by definition of f, we have

$$f(a, b) = a,\ f(c, d) = c.$$

Now $f[(a, b) + (c, d)] = f[a + c, b + d]$

$$= a + c$$

$$= f(a, b) + f(c, d)$$

Also, obviously f is onto G', Therefore f is a homomorphism of G onto G'.

Example 11:

If H be a normal subgroup of a finite group G, then prove that

$$o(G\,|\,H) = \frac{o(G)}{o(H)}$$

Solution:

o (G|H) = number of distinct right (or left) cosets of H in G, as G|H is the collection of all right (or left) cosets of H in G.

$$= \frac{\text{number of distinct elements in G}}{\text{number of distinct elements in H}}$$

$$= \frac{o(G)}{o(H)}, \text{ by Lagrange's theorem.}$$

Example 12:

Show that every quotient group of a cyclic group is cyclic but not conversely.

Solution:

Let H be a sub-group of a cyclic group G. Then H is also cyclic.

$\because$ every cyclic group is abelian

$\therefore$ H is a normal sub-group is G.

Let a be a generator of G and a^n be any element of G, where n is some integer.

Then Ha^n is any element of G|H.

Also, it can be proved easily that $(Ha)^n = Ha^n$, for every integer n.

$\therefore$ G|H is cyclic and its generator is Ha.

Its converse is not true. For example, if P_3 and A_3 be the symmetric and alternating groups on the three symbols a, b, c, then the quotient group $P_3|A_3$ is cyclic, whereas P_3 is not.

Example 13:

Show that every quotient group of an abelian group is abelian but its converse is not true.

Solution:

Let H be a sub-group of an abelian group G, so H is a normal sub-group of G.

Let a, b $\in$ G be arbitrary, then Ha, Hb are any two elements of the quotient group G|H.

Then we have

$$(Ha).(Hb) = Hab = Hba \qquad \because ab = ba \text{ if G is abelian.}$$
$$= (Hb).(Ha)$$

$\therefore$ G|H is abelian.

Its converse is not true. For example if P_3 and A_3 be the symmetric and alternating groups on the three symbols a, b, c, then the quotient group $P_3|A_3$ being of order 2 is abelian whereas P_3 is not.

Example 14:

Show that two right cosets Ha, Hb are distinct if and only if the two left cosets a^{-1} H, b^{-1} H are distinct.

Solution:

Suppose Ha and Hb are distinct. We have to prove that a^{-1} H, b^{-1} H are distinct.

If a^{-1} H, b^{-1} H are not distinct, then they are equal, i.e.,

$$a^{-1}H = b^{-1}H.$$

Now, $\quad a^{-1}H = b^{-1} H \Rightarrow a^{-1} \in b^{-1} H \Rightarrow ba^{-1} \in H.$

$\Rightarrow \quad (ba^{-1})^{-1} \in H \Rightarrow ab^{-1} \in H$

$\Rightarrow \quad a \in Hb \Rightarrow Ha = Hb.$

But this is a contradiction, since Ha and Hb are distinct.

It follows that a^{-1} H, b^{-1} H are not equal, i.e., they are distinct. The converse follows in the same manner,

Example 15(a):

Show that a homomorphism for a simple group is either trival or one-to one.

Solution:

Let G be a simple group and f be a homomorphism of G into another group G'. Then Kernel f is a normal subgroup of G. But the only normal subgroup of the simple group G is {e}. *Hence* either K = G or K = {e}. If K = G, the f-image of each element of G is the *identity* of G' and as such the homomorphism f is trivial one. If K = {e} the homomorphism f is one-to-one.

Examples 15(b):

Use Lagrange's theorem to prove that a finite group can not be expressed as the union of two of its proper sub-groups.

Solution:

Let G be a finite group of order n. Suppose G is the union of two of its proper sub-groups H and K..

Since e $\in$ both H and K and G = H $\cup$ K, therefore atleast one of H and K (say, H) must contain more than half the elements of G. Let o (H) = p. Then $n/2 < p < n$, since H is a proper sub-group of G.

Since $n/2 < p < n$ therefore p cannot be a divisor of n. This contradicts Lagrange's theorem which states that the order of each sub-group of a finite group is a divisor of the order of the group.

Hence our initial assumption is wrong and so a finite group cannot be expressed as the union of two of its proper sub-groups.

Example 15(c):

If H is a sub-group of G, and T is a subset defined as $T = \{X \in G;\ HX = XH\}$, then prove that T is sub-group of G.

Solution:

Let $X_1, X_2 \in T$, so that $HX_1 = X_1 H$ and $HX_2 = X_2H$.

We will first show that $X_2^{-1} \in T$. We see that

$$HX_2 = X_2H \Rightarrow X_2^{-1}(HX_2)X_2^{-1} = X_2^{-1}(X_2H)X_2^{-1}$$

$$\Rightarrow \quad X_2^{-1}\,H\,(X_2X_2^{-1}) = (X_2^{-1}\,X_2)\,H\,X_2^{-1}$$

$$\Rightarrow \quad X_2^{-1}H = HX_2^{-1}$$

Therefore $\quad X_2^{-1} \in T$.

Now, to show that $X_1X_2^{-1} \in T$, we see that

$$\begin{aligned} H\,(X_1X_2^{-1}) &= (HX_1)X_2^{-1} = (X_1H)X_2^{-1} \\ &= X_1(HX_2^{-1}) = X_1(X_2^{-1}H) \\ &= (X_1X_2^{-1})H \end{aligned}$$

Therefore $X_1X_2^{-1} \in T$.

Hence T is a sub-group.

Example 16:

Show that the set of the inverses of the elements of a right coset is a left coset, i.e., $(Ha)^{-1} = a^{-1}H$.

Solution:

Let Ha be a right coset of H in G where $a \in G$.

Now let $h\,a \in Ha$ where $h \in H$.

Since H is a sub-group.

$$h \in H \Rightarrow h^{-1} \in H$$

$$\therefore \quad a^{-1}h^{-1} \in a^{-1}H. \qquad [\because (ha)^{-1} = a^{-1}h^{-1}]$$

Thus the inverses of all the elements of Ha belong to the left coset $a^{-1}H$

Therefore $(Ha)^{-1} \not\subseteq a^{-1}H$.

Conversely, let $a^{-1}h$ be any element of $a^{-1}H$.

Then $a^{-1}h = a^{-1}(h^{-1})^{-1} = (h^{-1}a)^{-1}(Ha)^{-1}$

Since $h^{-1} \in H$ and hence $h^{-1}a \in Ha$.

Therefore every element of $a^{-1}H$ belongs to the set of inverses of the elements of Ha.

$$\therefore \quad a^{-1}H \subseteq (Ha)^{-1}$$

$$\text{Hence,} \quad (Ha)^{-1} = a^{-1}H.$$

Example 17:

Prove that the sets H and H^x are of the same order.

Solution:

We define H^x as follows :

$$H^x = x^{-1}Hx$$

Thus the elements of H^x are of the form $x^{-1}hx$, for some x and all $h \in H$.

Then $\quad h \Rightarrow x^{-1}hx$

Conversely

$$xhx^{-1} = x(x^{-1}hx)x^{-1}$$
$$= (xx^{-1})h(xx^{-1})$$
$$= h$$

Hence the correspondence between H and H^x is one-one.

Hence H and H^x have the same number of elements i.e., their order is the same.

Example 18(a):

Show that for any group G and an element $a \in G$, the subset $N(a) = [x \in G | xa = ax\}$ forms a subgroup of G.

Solution:

Clearly N (a) is non-empty as $e \in G$ and $e\, a = a = a\, e$, i.e.,

$$e \in N(a)$$

Let $\quad x_1, x_2 \in N(a)$ so that

$$ax_1 = x_1a \text{ and } ax_2 = x_2a \quad \text{...(i)}$$

Now $x_2 a = a x_2 \to x_2^{-1}(x_2 a)\ x_2^{-1}$

$= x_2^{-1}(a x_2)\ x_2^{-1}$

$\Rightarrow (x_2^{-1} x_2) a\ x_2^{-1} = (x_2 x_2^{-1}).$

$= a x_2^{-1} = x_2^{-1} a$ ...(ii)

Above implies that x_2^{-1} also commutes with $a \in G$ and as such x_2^{-1} also belongs to N(a).

Now $(x_1 x_2^{-1})\ a = x_1\ (x_2^{-1} a) = x_1\ (a x_2^{-1})$ by (ii)

$= (x_1\ a) x_2^{-1} = (a x_1)\ (x_2^{-1})$

$\because$ $a x_1 = x_1 a$ by (i)

$\therefore$ $(x_1 x_2^{-1})\ a = a\ (x_1 x_2^{-1})$

Hence $x_1\ x_2^{-1}$ also commutes with a and as such it belongs to N(a).

Thus $x_1, x_2 \in N(a) \Rightarrow x_1 x_2^{-1} \in N(a)$ which therefore is a sub-group.

Example 18(b):

H, K are two sub-groups of a group G, then HK is a sub-group of G iff HK = KH.

Solution:

(i) *Condition is necessary:* Let HK = KH then we have to prove that HK is a sub-group of G. For this we will prove $(HK)(HK)^{-1} = HK$.

Now, $(HK)(HK)^{-1} = (HK)(K^{-1}H^{-1})$ (reversal law)

$= H(KK^{-1})H^{-1}$ (associativity)

$= (HK)H^{-1}$

[$\because$ K is a sub-group $\therefore KK^{-1} = K$]

$= (KH)H^{-1}$ [$\therefore$ HK = KH]

$= K(HH^{-1})$

$= KH$ [$\because$ H is a sub-group $\therefore HH^{-1} = H$]

$\therefore$ HK = KH $\Rightarrow$ HK is a sub-group.

(ii) **Condition is sufficient :** Suppose that HK is a sub-group.

Then $(HK)^{-1} = HK \Rightarrow K^{-1}H^{-1} = HK$

$\Rightarrow$ KH = HK [$\because$ H is a subgroup $\Rightarrow H^{-1} = H$ and K is a subgroup $\Rightarrow K^{-1} = K$]

Hence the result.

Example 19:

$G = \{a, a^2, a^3, a^4 = 1\}$, $o(G) = 4$, $H = \{1, a^2\}$ is a sub-group of G. Find all the cosets of H in G and prove that G is equal to union of all these cosets and also establish that any two cosets are either disjoint or identical.

Solution:

We have

$$1.\ H = \{1, a^2] = H$$

$$a.\ H = \{a, a^3\},$$

$$a^2.\ H = \{a^2, a^4\} = \{a^2, 1\} = H \qquad [\because a^4 = 1]$$

$$a^3.\ H = \{a^3, a^5\} = \{a^3, a) = \{a, a^3\} = a.\ H \qquad [\because a^4 = 1]$$

Thus there are only two *distinct* cosets namely H and a H which are disjoint.

Also H is identical with a^2H and aH is identical with a^3H.

Again
$$H \cup a\ H = \{1, a^2\} \cup \{a, a^3\}$$
$$= \{1, a, a^2, a^3\} = G.$$

Example 20:

Let $G = P_3$, be the symmetric group on three symbols a, b, c and let

$$H = \left\{\begin{pmatrix} a & b & c \\ a & b & c \end{pmatrix}, \begin{pmatrix} a & b & c \\ a & c & b \end{pmatrix}\right\}$$ *be a sub-group of G, find the values of Hf*

and fH where $f = \begin{pmatrix} a & b & c \\ c & b & a \end{pmatrix} \in G$

Solution:

$$Hf = \left\{\begin{pmatrix} a & b & c \\ a & b & c \end{pmatrix}\begin{pmatrix} a & b & c \\ c & b & a \end{pmatrix}, \begin{pmatrix} a & b & c \\ a & c & b \end{pmatrix}\begin{pmatrix} a & b & c \\ c & b & a \end{pmatrix}\right\}$$

$$= \left\{\begin{pmatrix} a & b & c \\ c & b & a \end{pmatrix}, \begin{pmatrix} a & b & c \\ c & a & b \end{pmatrix}\right\}$$

$$fH = \left\{\begin{pmatrix} a & b & c \\ c & b & a \end{pmatrix}\begin{pmatrix} a & b & c \\ a & b & c \end{pmatrix}, \begin{pmatrix} a & b & c \\ c & b & a \end{pmatrix}\begin{pmatrix} a & b & c \\ a & c & b \end{pmatrix}\right\}$$

$$= \left\{ \begin{pmatrix} a & b & c \\ c & b & a \end{pmatrix}, \begin{pmatrix} a & b & c \\ b & c & a \end{pmatrix} \right\}$$

Example 21(a):

If S = {0, 2, 4} then show that (S,⁺ 6) a sub-group of the group (I₊⁶6)

Solution:

We know I_6, = {0, 1, 2, 3, 4, 5}

∴ $S \subseteq I_6$ and is non-empty.

Also, for the system (S, +6) the composition table is

$^+6$	0	2	4
0	0	2	4
2	2	4	0
4	4	0	2

From the above table we conclude that—

(i) **The closure property** is satisfied since each entry in the table is an element of S.

(ii) $0 +_6 (2 +_6 4) = 0 +_6 0,$ $\quad \because 2 +_6 4 = 0$ from the table

$= 0$, from the table

and $(0 +_6 2) +_6 4 = 2 +_6 4$ $\quad \because 0 +_6 2 = 2$ from the table

$= 0$ from the table

Hence we find $0 +_6 (2 +_6 4) = (0 +_6 2) +_6 4$

Similarly, we can show that

$$a +_6 (b +_6 c) = (a +_6 b) +_6 c, \ \forall\ a, b, c \in S.$$

Hence the *associative proeprty* is satisfied.

(iii) From the table if is evident that

$$a +_6 0 = a = 0 +_6 a, \ \forall\ a, \in S.$$

i.e., 0 is the *identity element* is S for the operation.

(iv) From the table we observe that the inverse of 0, 2, and 4 are 0, 4 and 2 respectively.

Thus all the group postuletes are satisfied and so the set S forms a group under addition modulo 6.

Hence (S, ⁺6) is a sub-group of the group $(I_6, +_6)$.

Example 21(b):

Find all the sub-groups of a cyclic group of order 1 2.

Solution:

We know that the integral divisors of 12 are 1, 2, 3, 4, 6, 12. Now, there exists one and only one sub-group of each of these orders.

Let a be the generators of the group and m be a divisor of 12. Then there exists one and only one element G whose order is m, i.e., $a^{12/m}$.

∴ All the elements of order 1, 2, 3, 4, 5, 6, 12, will give sub-groups.

∴ $(a^{12}) = \{e\}$, (a^6), (a^4), (a^3), (a^2), (a) are the required sub-groups.

Example 21(c):

(i) can an abelian group have a non-abelian sub-group ?

(ii) can a non-abelian group have an abelian sub-group ?

(iii) can a non-abelian group have a non-abelian sub-group ?

Solution:

(i) Every sub-group of an abelian group is abelian. If G is an abelian group and H is a sub-group of G, then the operation on H is commutative because it is already commutative in G and H is a subset of G. Hence an abelian group have a non-abelian sub-group.

(ii) A non-abelian group can have an abelian sub-group. For example, the symmetric group P_3 of permutations of degree 3 is non-abelian while its sub-group A_3 is abelian.

(iii) A non-abelian group can have a non-abelian sub-group. For example P_4 is a non-abelian group and its sub group A_4 is also non-abelian.

Example 22:

How many elements of the group $G = [a, a^2, a^3, a^4, a^5, a^6 = e]$ can be used as generators of the group ?

Solution:

From the definition of the generator of a group, it is evident that a is a generator of the group **G**.

Also, by the property of cyclic group, a^p is a generator of **G**, provided p is prime to n, i.e., p is prime to 6 (∵ Here n = 6). Therefore, here p = 5.

Hence a and a^5 can be used as generators of **G**.

Example 23:

Show that the set G = {1, 2, 3, 4, 5, 6} (mod. 7) with respect to multiplication is a cyclic group.

Solution:

Here we find that:

Closure property : 1.2 = 2, 1.3 = 3, 1.4 = 4, 1.5 = 5, 1.6 = 6, 2.3 = 6

2.4 = 8 ≡ 1 (Mod 7); 2.5 = 10 ≡ 3 (Mod 7);

2.6 = 12 ≡ 5 (Mod 7); 3.4 = 12 ≡ 5 (Mod 7);

3.5 = 15 ≡ 1 (Mod 7); (3.6) = 18 ≡ 4 (Mod 7);

4.5 = 20 ≡ 6 (Mod 7); 4.6 = 24 ≡ 3 (Mod 7);

5.6 = 30 ≡ 2 (Mod 7)

∴ The closure property is satisfied

Associativity : 2.(3.4) = 2.5 [∵ 3.4 = 12 = 5 (Mod 7)]

= 3 [∵ 2.5 = 10 = 3 (Mod 7)]

And (2.3).4 = (6).4 = 3, [∵ 6.4 = 24 = 3 (Mod 7)]

∴ (2.3).4 = 2. (3.4)

We can similarly prove other results taking other elements. Hence the associative law holds.

Identity Element : The identity element exists and is 1.

Inverse : As in closure property, the inverse of each element exists and the inverse of 1, 2, 3, 4, 5 and 6 are 1, 4, 5, 2, 3, 6, respectively.

Hence (**G**, 7) is a group. Now we are to prove that it is cyclic for this let there exist an element a ∈ G such that o (a) = o (**G**) = 6, then the group **G** will be a cyclic group and a will be its generator.

Here we find that $3^1 = 3$; $3^2 = 3.3 = 9 \equiv 2(\text{Mod } 7)$;

$3^3 = 3^2.3 = 2.3$,

∴ $3^2 = 2 \text{ (mod 7) or } 3^3 = 6$

$= o\ (G)$ ...(i)

$3^4 = 3^3 = 6.3$ [∵ $3^4 \equiv 4 \text{ (mod 7)}$

i.e., $3^4 = 18 \equiv 4 \text{ (mod 7)}$

$3^5 = 3^4.3 = 4.3$, [∵ $3^4 \equiv 4 \text{ (mod 7)}$

or $3^5 = 12 \equiv (5 \text{ mod } 7)$ and $3^6 = 3^5.3 = 15 \equiv 1 \text{ (mod 7)}$

From (i) we observe that o (3) = 6 = o (G) and so 3 is a generator of G we also have found that

$$36 = 1,\ 3^2 = 2,\ 3^1 = 3,\ 3^4 = 4,\ 3^5 = 5,\ 3^3 = 6$$

Hence (G, 7) is a cyclic group whose generator is 3.

Example 24(a):

Show that the set of non-zero residue classes modulo 5 is a cyclic group under multiplication modulo 5.

Solution:

Let G=(1}, {2}, {3}, {4}} be the set of non-zero residue classes modulo 5.

It can easily be shown that this forms a multiplicative group and (1) is its identity element.

Now in order to prove that this group is cyclic we are to prove that there exists an element a ∈ **G** such that o (a) = o (**G**) = 4 , the number of elements in **G** being 4.

Evidently $2 = \{2\};\ \{2\}\ \{2\} = \{4\};\ \{2\}.\{2\}.\{2\} = \{4\}.\{2\}.$

Since $\{2\}.\{2\} = \{4\}$

$= \{3\}$, since $4.2 = 8 = 3 \pmod 5$

$\{2\}\ .\ \{2\}\ .\ \{2\}\ .\ \{2\} = (3).\ \{2\}\ \because\ \{2\}.\ \{2\}.\ \{2\} = \{3\}$

$= \{1\}$. since $3.2 = 6 \equiv 1 \pmod 5$

$\therefore\ o(\{2\}) = 4 = o(G)$

Hence the given set G forms a cyclic group.

Example 24(b):

Prove that the set of nth roots of unity is a finite abelian cyclic group with multiplicative composition.

Solution:

We know that

$$1 = \cos 0 + i \sin 0$$

$$\therefore\ 1^{1/n} = (\cos^{2r\pi} + i\ \sin^{2r\pi})^{1/n}$$

$$= \left(\cos\frac{2r\pi}{n} + i\,\frac{2r\pi}{n}\right)$$

where r = 0, 1, 2, ..., (n – 1)

$= e^{2r\pi i/n}$

Let G denotes the set of nth roots of unity, then

$$G = \{e^{2r\pi i/n} : = 0, 1, 2, ..., (n-1)\}$$

Now, let us verify the group axioms :

(i) Closure axiom : Let

$$e\frac{2\pi r_1 i}{n}, e\frac{2\pi r_2 i}{n} \in G$$

then $$e\frac{2\pi r_1 i}{n} . e\frac{2\pi r_2 i}{n} = e\frac{2\pi(r_1+r_2)i}{n} \in G$$

$$r_1 + r_2 \leq n-1$$

If $r_1 + r_2 > n-1$, let us suppose $r_1 + r_2 = n + k$ where $k \leq n-2$ then

$$e\frac{2\pi r_1 i}{n} . e\frac{2\pi r_1 i}{n} = e\frac{2\pi(r_1+r_2)i}{n}$$

$$e\frac{2\pi}{n}(n+k)i = e^{2\pi i}.e\frac{2\pi}{n}ki$$

$$1.e\frac{2\pi ki}{n} = e\frac{2\pi ki}{n} \in G.$$

Hence the set G is closed under multiplication.

(ii) Associative axiom: The multiplication for complex number is always associative.

(iii) Identity axiom: $\cos 0 + i \sin 0 = 1 = e^{2\pi i 0/n}$ is an identity element is G

(iv) Inverse axiom: The inverse of $e\frac{2\pi ri}{n}$ is $e\frac{2\pi(n-r)i}{n}$ because

$$e\frac{2\pi ri}{n} \text{ is } e\frac{2\pi(n-r)i}{n}$$

$$= e\frac{2\pi ri}{n} \text{ is } e^{2\pi ri}.e\frac{-2\pi ri}{n}$$

$$-1.e\left(\frac{2\pi ri}{n} - \frac{2\pi ri}{n}\right) - e\frac{2\pi(r_1+r_2)i}{n} = e^0 = 1$$

(v) Commutative axiom: Since multiplication of complex numbers is commutative, G is a multiplicative abelian group.

Also, the number of elements in G is n, i.e., finite and all the elements are (generated by $e^{2\pi i/n}$ Hence G is a finite abelian cyclic group.

Example 25:

Show that there exist only two generators of any infinite cyclic group.

Solution:

Let G is any cyclic group whose generator is a, i.e., G = {a}.

Then as any integral power of a can be expressed as some integral power of a^{-1} (i.e., the inverse of a) and vice-versa.

So $a^{-1}$is also a generator of G. Thus we have a and a^{-1} as the two generators of G. If possible let a^p be another generator of G besides these two, then $a = (a^p)^m$, for some $m \in I$, the set of integers.

$$aa^{-1} = (a^p)^m\, a^{-1}$$

$\Rightarrow \quad e = a^{pm-1}, \qquad [\because aa^{-1}e]$

$\Rightarrow \quad a^o = a^{pm-1}$

$\Rightarrow$ either $p = m = 1$ or $p = m = -1$

$\Rightarrow$ a^p is a generator of G only if $p = 1$ or -1

$\therefore$ a and a^{-1} are only two generators of G.

Example 26:

Prove that any group G is order 3 is cyclic.

Solution:

Since G is of order 3, it must have exactly two distinct elements other than identity element e. Let these elements be a and b.

Since G is a group, closure law must hold and hence

$$ab \in G \qquad [\therefore ab = e \text{ or } a\,b = a]$$

or $$ab = b$$

If ab = a then b = e which is contradictory to the hypothesis.

Similarly a b = b also leads to the contradiction because ab = b

$$\Rightarrow a = e.$$

Hence essentially

$$ab = e \qquad ...(i)$$

Also $a^2 \in G$ hence $a^2 = e$ or $a^2 = a$ or $a^2 = b$

But $a^2 = ab \Rightarrow a.a = ab \Rightarrow a = b$ (by left cancellation law)

which again contradicts the hypothesis.

Therefore the only possibility is

$$a^2 = b$$

Thus $\quad G = \{e, a, a^2\}$

But $\quad a^3 = a.a^2 = ab = e$

Finally we can write $G = \{a, a^2, a^3, (= e)\}$

Hence G is a cyclic group whose generator is a..

Example 27:

Find the order of each element of the multiplicative group G, where

$$G = \{1, -1, i, -i\}$$

Solution:

Since 1 is the identity element, its order is 1.

Now $\quad (-1)^1 = -1, (-1)^2 = (-1)(-1) = 1$

Hence order of -1 is 2.

Again $\quad i^1 = i, i^2 = -1, i^3 = -i, i^4 = 1$

Therefore order of i is 4.

Similarly, $\quad (-i)^1 = -i, (-i)^2 = i^2 = -1\ (-i)^3 = +i, (-i)^4 = -1$

Hence order of $-$ i is 4.

Example 28(a):

Find the order of each element in the following multiplicative group G.

$$G = \{a, a^2, a^3, a^4, a^5, a^6 = e\}$$

Solution:

The order of a is a 6 as $a^6 = e$

The order of a^2 is 3 as $(a^2)^3 = a^6 = e$

The order of a^3 is 2 as $(a^3)^2 = a^6 = e$

The order of a^4 is 3 as $(a^4)^3 = a^{12} = (a^6)^2 = a^2 = e$

The order of a^5 is 6 as $(a^5)^6 = a^{30} = (a^6)^5 = e^5 = e$

and the order of a^6 is 1 as $(a^6)^1 = a^6 = e$.

Example 28(b):

If the elements a, b of a group commute and o (a) = m, o (b) = n, where (m, n) = 1, prove that o(ab) = mn.

Solution:

We are given that the two elements a and b of a group commute and o (a) = m, o (b) = n with (m, n) = 1

We have to prove that $o(ab) = mn$

Let $o(ab) = k.$

Since a and b commute, we have

$$(ab)^{mn} = a^{mn} b^{mn} = (a^m)^n (b^n)^m = e^n e^m$$
$$= ee = e$$

Again $(ab)^k = a^k b^k$ i.e., $e = a^k b^k$ [$\because$ $o(ab) = k$ so that $(ab)^k = e$]

But $a^k b^k = e \; a^k = (b^k)^{-1}$

$\Rightarrow$ $o(a^k) = o\{(b^k)^{-1}\}$

$\Rightarrow$ $o(a^k) = o(b^k)$

We have $o(a^k) \mid o(a)$ and $o(b^k) \mid o(b)$, i.e., $o(a^k) \mid m$ and $o(b^k) \mid n$.

Since $o(a^k) = o(b^k)$ we see that either $o(a^k$ or $o(b^k)$ divides both m and n and so it divides their H. C. F $(m, n) = 1$. In other words, $o(a^k) = o(b^k) = 1$, so that $a^k = e$ and $b^k = e$. Hence, we have $m|k$ and $n|k$. Together with $(m, n) = 1$, this gives $mn|k$ and combining this with $k| mn$, we obtain $k = mn$, i.e., $o(ab) = mn$.

Example 29:

If the elements A, B and A B of a finite group are each of order 2, prove that A B = B A

Solution:

Given that $A^2 = e$, $B^2 = e$ and $(AB)^2 = e$, where e is the identity element.

Now, $(AB)^2 = e$

$\Rightarrow$ $(AB)(AB) = e$

$\Rightarrow$ $A(AB)(AB) = Ae$

$\Rightarrow$ $A^2(BA)B = A$

$\Rightarrow$ $\{e(BA)B\}B = AB$

$\Rightarrow$ $(BA)B^2 = AB$

$\Rightarrow$ $(BA)e = AB$

$\Rightarrow$ $BA = AB$

Thus $AB = BA.$

Example 30:

In a group if $ba = a^m b^n$., prove that the elements $a^m b^{n-2}$, $a^{m-2} b^m$ and a^{b-1} have the same order.

Solution:

We can write

$$a^m b^{n-2} = a^m (b^n ab^{-2})$$
$$= (a^m b^n)b^{-2} = (ba)b^{-2} \qquad (\because ba = a^m b^n)$$
$$= b(ab^{-1})b^{-1}$$

Hence $\quad o(a^m b^{n-2}) = o[b(ab^{-1})b^{-1}]$

But we know that $o[b(ab^{-1})b^{-1}] = o(ab^{-1})$

$$o(a^m b^{n-2}) = o(ab^{-1})$$

Again, writing $\quad a^{m-2} b^n = a^{-2}(a^m b^n)$

$$= a^{-2}(ba)$$
$$= a^{-2}(ba^{-1})a^2$$
$$= (a^2)^{-1}(ba^{-1})a^2$$

We have $\quad o(a^{m-2} b^n) = o(ba^{-1})$

$$= o[(ab^{-1})^{-1}] = o(ab^{-1})$$

$\therefore \quad o(a^m b^{n-2}) = o(a^{m-2} b^n) = o(ab^{-1})$

Example 31:

If G is a group such that $(ab)^m = a^m b^m$ for three consecutive integers m for all a, b $\in$ G, show that G is abelian.

Solution:

Let a, b be any two elements of **G**. Suppose m, m + 1, m + 2 are three consecutive integers such that

$$(ab)^m = a^m b^m, (ab)^{m+1} = a^{m+1} b^{m+1},$$

and $\quad (ab)^{m+2} = a^{m+2} b^{m+2}$

We have $\quad (ab)^{m+2} = (ab)^{m+1}(ab)$

$\Rightarrow \quad aa^{m+1}b^{m+1} b = aa^m b^m bab$

$$a^{m+1} b^{m+1} = a^m b^m (ba)$$

[by left and right cancellation laws]

$\Rightarrow \quad (ab)^{m+1} = (ab)^m (ba)$

$\Rightarrow \quad (ab)^{m+1}(ab) = (ab)^m (ba)$

$\Rightarrow \quad ab = ba \qquad$ [by left cancellation law]

$\Rightarrow \quad$ **G** is abelian.

Example 32:

Find the order of elements of the group {0, 1, 2, 3,4, 5} the composition being addition modulo 6.

Solution:

Since 0 is the identity element,

Hence, o (0) = 1.

Now, $(1)^1 = 1, 1^2 + 1 = {}_61; 1^3 = 1 + {}_61 = 1 +_6 2 = 3,$

$1^4 = 1 + {}^63 = 4,\ 1^5 = 1 + {}_64 = 5,\ 1^6 + {}_61 = 1 +_6 5 = 1 + {}_65 = 0$

∴ o (1) = 6

Again $2^1 = 2, 2^2 = 2 +_6 2 = 4, 2^3 = 2 +_6 2^2 = 2 +_2 +_6 4 = 0$ (Identity element)

∴ o (2) = 3

Further $3^1 = 3\ ,\ 3^2\ 3 +_6 3 = 0$ (i.e., identity element)

∴ o (3) = 2

Similarly, we can find o (4) = 3 , o (5) = 6.

Example 33:

In the additive group of integers, show that the order of every element except 0 is infinite.

Solution:

0 is the identity element. Hence o (0) = 1.

Now, 1 ∈ 1. We have 1 (1) = 1, 2(1) = 2, 3 (1) = 3 and so on. Thus there exists no positive integer n such that n (1) = 0 (identity element). Therefore infinite.

Example 34:

Write the following permutations as the product of disjoint cycles.

(a) $f = \begin{pmatrix} 1 & 2 & 3 & 4 & 5 & 6 & 7 & 8 & 9 \\ 2 & 3 & 4 & 5 & 1 & 6 & 7 & 8 & 9 \end{pmatrix}$

(b) $g = \begin{pmatrix} 1 & 2 & 3 & 4 & 5 & 6 \\ 6 & 5 & 4 & 3 & 1 & 2 \end{pmatrix}$

Solution:

(a) We have (f) = (6) (7) (1 2 3 4 5) (8 9)

or f = (1 2 3 4 5) (8 9), omitting cycles of length 1 as they represent identity permutation.

(b) We have g = (1 6 2 5) (3 4).

Example 35:

Show that the set of all integral multiples of an integer m forms a sub-group of the additive group of all integers.

Solution:

Let (I, +) be the additive group of all integers and

$$S = \{\lambda\, m : m \text{ is a fixed integer and } \lambda \in I\}$$
$$= \{\ \ldots\ldots, -3m, -2m, -m, 0, m, 2m, 3m, \ldots\ldots\}$$

Then we find that—

(i) If am and bm be two elements of S, where a, b $\in$ I, then we get

am + bm = (a + b) m $\in$ S, $\because$ a + b $\in$ I, i.e. the closure property is satisfied.

(ii) $\forall$ a, b, c, $\in$ I, we get

$$am + (bm + cm) = am + (b + c)\, m$$
$$= (a + b + c)m$$

and
$$(am + bm) + cm = (a + b)\, m + cm$$
$$= (a + b + c)\, m$$

Hence the operation of addition is *associative* in S.

(iii) Since $0 + \lambda m = \lambda m = \lambda m + 0,\ \forall\ \lambda m \in S$

So 0 is the identity of S for addition.

(iv) Also, we find $-\lambda m \in S$ is the inverse of $\lambda m \in S$, since

$$(-\lambda m) + \lambda m = 0 = (\lambda m) + (\lambda m)$$

Hence the inverse of each element of S exists.

Thus all the group postulates are satisfied and so the set S forms a group under addition.

Hence (S, +) is a sub-group of the group (I, +).

Example 36:

Find $(a_i\ a_2)\ (a_1\ a_3)$

Solution:

$$(a_1a_2)(a_1a_3) = \begin{pmatrix} a_1 & a_2 & a_3 \\ a_2 & a_1 & a_3 \end{pmatrix}\begin{pmatrix} a_1 & a_2 & a_3 \\ a_3 & a_2 & a_1 \end{pmatrix}$$

$$= \begin{pmatrix} a_1 & a_2 & a_3 \\ a_2 & a_1 & a_3 \end{pmatrix}\begin{pmatrix} a_2 & a_1 & a_3 \\ a_2 & a_3 & a_1 \end{pmatrix}$$

$$= \begin{pmatrix} a_1 & a_2 & a_3 \\ a_2 & a_3 & a_1 \end{pmatrix} = (a_1 \quad a_2 \quad a_3)$$

Example 37:

Determine which of the following are even permutations :

(a) = f (1 2 3 (1 2)

(b) = g (1 2 3 4 5) (1 2 3) (4 5)

(c) = h (1 2) (1 3) (1 4) (2 5).

Solution:

(a) We can write f = (1 2) (1 3) (1 2). The number of transpositions is 3, i.e., odd, Hence f is odd permutations.

(b) We have g = (1 2) (1 3) (1 4) (1 5) (1 2) (1 3) (4 5)

The number of transpositions is 7, i.e., odd.

Hence g is an odd permutation

(c) h = (1 2) (1 3) (1 4) (2 5).

The number of transpositions is 4, i.e., even. Hence h is an even permutation.

Example 38(a):

Write down all the permutations on four symbols 1, 2, 3, 4. Which of these permutations are even ?

Solution:

There will 4 ! : i.e., 2 4 permutations of degree 4. If P_4 is the set of the all these permutation, then

P_4 = (1) (1 2) (1 3) (1 4) (2 3), (2 4) (3 4) (1 2 3)(1 3 2) (1 2 4), (1 4 2), (1 3 4), (1 4 3), (2 3 4), (2 4 3), (1 2), (3 4), (2 3), (1 4). (3 1) (2 4), (1234) (1 2 4 3), (1 3 4), (1 3 2 4), (1 3 4 2), (1 4 2 3), (1 4 3 2).

If A_4 is the set of all even permutations of degree 4, then A_4 will have 1/2 × 4! i.e., 12 elements.

Thus A_4 = (1), (1 2 3), (1 3 2), (1 2 4), (1 3 4), (1 4 3), (2 3 4), (2 4 3), (1 2) (3 4), (2 3), (14), (31), (2 4).

Example 38(b):

What integral power of $f = \begin{pmatrix} 1 & 2 & 3 & 4 \\ 1 & 3 & 4 & 2 \end{pmatrix}$ *be multiplied by itself to product.* $\begin{pmatrix} 1 & 2 & 3 & 4 \\ 1 & 2 & 3 & 4 \end{pmatrix}$

Solution:

$$\begin{pmatrix}1 & 2 & 3 & 4\\1 & 3 & 4 & 2\end{pmatrix}\begin{pmatrix}1 & 2 & 3 & 4\\1 & 3 & 4 & 2\end{pmatrix}$$

$$=\begin{pmatrix}1 & 2 & 3 & 4\\1 & 3 & 4 & 2\end{pmatrix}\begin{pmatrix}1 & 3 & 4 & 2\\1 & 4 & 2 & 3\end{pmatrix}$$

$$=\begin{pmatrix}1 & 2 & 3 & 4\\1 & 4 & 2 & 3\end{pmatrix}$$

Again, $$\begin{pmatrix}1 & 2 & 3 & 4\\1 & 4 & 2 & 3\end{pmatrix}\begin{pmatrix}1 & 2 & 3 & 4\\1 & 3 & 4 & 2\end{pmatrix}$$

$$=\begin{pmatrix}1 & 2 & 3 & 4\\1 & 4 & 2 & 3\end{pmatrix}\begin{pmatrix}1 & 2 & 2 & 3\\1 & 2 & 3 & 4\end{pmatrix}$$

$$=\begin{pmatrix}1 & 2 & 3 & 4\\1 & 2 & 3 & 4\end{pmatrix}$$

Hence the permutation f^2 will yield to the desired result.

Examples 39:

Show that the set of six transformations $f_1, f_2, f_3, f_4, f_5, f_6$ on the set of complex numbers defined by

$$f_1(z) = z,\ f_2(z) = 1/z,\ f_3(z) = 1 - z$$

$$f_4(z) = \left(\frac{z}{z-1}\right),\ f_5(z) = \frac{1}{1-z},\ f_6(z) = \frac{z-1}{z}.$$

forms a finite non-abelian group of order 6 with respect to the composite composition.

Solution:

First of all we prepare a composition table. Clearly identity element for the composition is the identity mapping f_1, calculations can be made in the following manner:

$$(f_2 \circ f_2)(z) = f_2(f_2(z)) = f_2(1/z) = z$$

$$= f_1(z) \text{ so that } f_2 \circ f_2 = f_1$$

$$(f_2 \circ f_3)(z) = f_2(f_3(z)) = f_2(1 - z) = 1/1 - z$$

$$= f_5(z),$$

so that $$f_2 \circ f_3 = f_5.$$

$$(f_2 \circ f_4)(z) = f_2(f_4(z)) = f_2\left(\frac{z}{z-1}\right) = \frac{z-1}{z}$$

$$= f_6(z)$$

so that $(f_2 \circ f_4) = f_6.$

$$(f_2 \circ f_5)(z) = f_2\left(\frac{1}{z-1}\right) = 1 - z = f_3(z)$$

$$(f_2 \circ f_6)(z) = f_2\left(\frac{z-1}{z}\right) = \frac{z}{z-1}\ f_4(z)$$

This completes the second row of the table. Similarly other entries can be filled up. These are left as an exercise for the students. We thus get the following composition table.

0	f_1	f_2	f_3	f_4	f_5	f_6
f_1	f_1	f_2	f_3	f_4	f_5	f_6
f_2	f_2	f_1	f_5	f_6	f_3	f_4
f_3	f_3	f_6	f_1	f_5	f_4	f_2
f_4	f_4	f_5	f_6	f_1	f_2	f_3
f_5	f_5	f_4	f_2	f_3	f_6	f_1
f_6	f_6	f_3	f_4	f_2	f_1	f_5

The following facts are evident from the table :

(i) Product of any two transformations of the set is again in the set so that closure axiom is satisfied.

(ii) The identity transformation f_1 is the *identity* element.

(iii) Every transformation has an *inverse.* Thus

$f_1^{-1} = f_1$, $f_2^{-1} = f_2$, $f_3^{-1} = f_3$, $f_4^{-1} = f_4$,

$f_5^{-1} = f_6$ and $f_6^{-1} = f_5$

(iv) The composition is *associative* since the composite composition is associative in the set of all mappings in general although it is tedious to verify it from table.

(v) The composition is not commutative since

$$f_2 \circ f_3 \neq f_3 \circ f_2 \text{ etc.}$$

Hence the set of six transformations forms a finite non-abelian group of order six with respect to the composite composition.

Example 40:

Verify the following :

(a) (1 2 3) (4 5) = (4 5) (1 2 3)

(b) (1 2 3) (2 3) ≠ (2 3) (1 2 3)

Hence deduce that AB = BA if A and B are disjoint cycles.

Solution:

(a) L.H.S. = (1 2 3) (4 5)

$$= \begin{pmatrix} 1 & 2 & 3 & 4 & 5 \\ 2 & 3 & 1 & 4 & 5 \end{pmatrix} \begin{pmatrix} 1 & 2 & 3 & 4 & 5 \\ 1 & 2 & 3 & 5 & 4 \end{pmatrix}$$

$$= \begin{pmatrix} 1 & 2 & 3 & 4 & 5 \\ 2 & 3 & 1 & 4 & 5 \end{pmatrix} \begin{pmatrix} 2 & 3 & 1 & 4 & 5 \\ 2 & 3 & 1 & 5 & 4 \end{pmatrix}$$

$$= \begin{pmatrix} 1 & 2 & 3 & 4 & 5 \\ 2 & 3 & 1 & 5 & 4 \end{pmatrix}$$

R.H.S. = (4 5) (1 2 3)

$$= \begin{pmatrix} 1 & 2 & 3 & 4 & 5 \\ 1 & 2 & 3 & 5 & 4 \end{pmatrix} \begin{pmatrix} 1 & 2 & 3 & 4 & 5 \\ 2 & 3 & 1 & 4 & 5 \end{pmatrix}$$

$$= \begin{pmatrix} 1 & 2 & 3 & 4 & 5 \\ 1 & 2 & 3 & 5 & 4 \end{pmatrix} \begin{pmatrix} 1 & 2 & 3 & 4 & 5 \\ 2 & 3 & 1 & 4 & 5 \end{pmatrix}$$

$$= \begin{pmatrix} 1 & 2 & 3 & 4 & 5 \\ 2 & 3 & 1 & 5 & 4 \end{pmatrix}$$

∴ L.H.S. = R.H.S.

(b) $(1\ 2\ 3)\ (2\ 3) = \begin{pmatrix} 1 & 2 & 3 \\ 2 & 3 & 1 \end{pmatrix} \begin{pmatrix} 1 & 2 & 3 \\ 1 & 3 & 2 \end{pmatrix}$

$$= \begin{pmatrix} 1 & 2 & 3 \\ 2 & 3 & 1 \end{pmatrix} \begin{pmatrix} 2 & 3 & 1 \\ 3 & 2 & 1 \end{pmatrix}$$

$$= \begin{pmatrix} 1 & 2 & 3 \\ 3 & 2 & 1 \end{pmatrix}$$

$$= (1, 3)$$

And $(2\ 3)\ (1\ 2\ 3) = \begin{pmatrix} 1 & 2 & 3 \\ 1 & 3 & 2 \end{pmatrix} \begin{pmatrix} 1 & 2 & 3 \\ 2 & 3 & 1 \end{pmatrix}$

$$= \begin{pmatrix} 1 & 2 & 3 \\ 1 & 3 & 2 \end{pmatrix} \begin{pmatrix} 1 & 3 & 2 \\ 2 & 1 & 3 \end{pmatrix}$$

$$= \begin{pmatrix} 1 & 2 & 3 \\ 2 & 1 & 3 \end{pmatrix}$$

$$= (1, 2)$$

Hence, (1 2 3) (2 3) ≠ (2 3) (1 2 3)

It is clear from above that AB = BA if A and B are disjoint cycles.

Example 41:

Prove that a symmetric set P_3, of all permutations of degree 3 defined on a set $A = \{1, 2, 3\}$ forms a finite group of order 6, the operation being composite of permutations.

Solution:

There shall be 3! distinct permutations, hence

$$P_3 = \left\{\begin{pmatrix}1&2&3\\1&2&3\end{pmatrix}, \begin{pmatrix}1&2&3\\1&3&2\end{pmatrix}, \begin{pmatrix}1&2&3\\2&1&3\end{pmatrix}, \begin{pmatrix}1&2&3\\2&3&1\end{pmatrix}, \begin{pmatrix}1&2&3\\3&1&2\end{pmatrix}, \begin{pmatrix}1&2&3\\3&2&1\end{pmatrix}\right\}$$

Let these elements be denoted by f_1, f_2, f_3, f_4, f_5, f_6, respectively. Now let us prepare a composition table

Product of Permutation	f_1	f_2	f_3	f_4	f_5	f_6
f_1	f_1	f_2	f_3	f_4	f_5	f_6
f_2	f_2	f_1	f_5	f_6	f_3	f_4
f_3	f_3	f_4	f_1	f_2	f_6	f_5
f_4	f_4	f_3	f_6	f_5	f_1	f_2
f_5	f_5	f_6	f_2	f_1	f_4	f_3
f_6	f_6	f_5	f_4	f_3	f_2	f_1

From this table we observe that:

(i) Since all the entries in the table belong to P_3 closure axiom is satisfied.

(ii) Composition of permutations is associative.

(iii) f_1 is identity element in P_3

(iv) Inverse of f_1, f_2, f_3, f_4, f_5, f_6, are f_1, f_2, f_3, f_4, and f_6 respectively.

Thus the symmetric set P_3 forms a finite group of order $3! = 6$.

Example 42(a):

Show that the four permutations I, (a b), (c d), (a b) (c d) on four symbols a, b, c, d from a finite abelian group with respect to the permutation multiplication.

Solution:

Let $I = f_1$, $(ab) = f_2$, $(cd) = f_3$, $(ab)(cd) = f_4$. To prepare the composition table, we observe that f_2 and f_3 are transpositions. Therefore $f_2 f_2 = f_1$, $f_3 f_3 = f_1$. Also f_2 and f_3 are disjoint cycles.

Therefore, $f_2 f_3 = f_3 f_2 = f_4$.

Further $f_2 f_4 = (ab)(ab)(cd)$

$= I(cd) = (cd) = f_3$

Similarly, $f_3 f_4 = (cd)(ab)(cd)$

$= (cd)(cd)(ab) = I(ab)$

$= (ab) = f_2$ [$\because$ (ab) (cd) =(cd) (ab)]

Also, $f_4 f_4 = (ab)(cd)(ab)(cd)$

$= (ab)(ab)(cd)(cd)$

$= (I)(I) = I = f_1$.

Similarly, making all other calculations, the composition table is

Product of Permutation	f_1	f_2	f_3	f_4
f_1	f_1	f_2	f_3	f_4
f_2	f_2	f_1	f_4	f_3
f_3	f_3	f_4	f_1	f_2
f_4	f_4	f_3	f_2	f_1

From the table it is clear that

(i) All the entries in the composition table are elements of the given set. Therefore, the closure axiom is satisfied.

(ii) f_1 is the identy element.

(iii) Each element possesses *inverse* In fact $f_1^{-1} = f_1$, $f_2^{-1} = f_2$, $= f_3^{-1} = f_3$, $f_4^{-1} = f_4$.

(iv) The composition is *commutative.*

(v) The multiplication of permutations is an *associative* composition. Hence, the given set is a finite abelian group of order 4 with respect to the permutation multiplication.

Example 42(b):

If $fg - \left\{\begin{pmatrix} 1 & 2 & 3 \\ 1 & 3 & 2 \end{pmatrix} \text{ and } g = \begin{pmatrix} 1 & 2 & 3 \\ 2 & 3 & 1 \end{pmatrix}\right\}$ *be two permutations of degree 3, Find fg and gf and show that fg ≠ gf.*

Solution:

$$fg = \left\{\begin{pmatrix} 1 & 2 & 3 \\ 1 & 3 & 2 \end{pmatrix}\begin{pmatrix} 1 & 2 & 3 \\ 2 & 3 & 1 \end{pmatrix}\right\}$$

$$=\left\{\begin{pmatrix}1&2&3\\1&3&2\end{pmatrix}\begin{pmatrix}1&2&3\\2&1&3\end{pmatrix}=\begin{pmatrix}1&2&3\\2&1&3\end{pmatrix}\right\}$$

and $$gf=\begin{pmatrix}1&2&3\\2&3&1\end{pmatrix}\begin{pmatrix}1&2&3\\1&3&2\end{pmatrix}$$

$$=\begin{pmatrix}1&2&3\\2&3&1\end{pmatrix}\begin{pmatrix}2&3&1\\3&2&1\end{pmatrix}$$

$$=\begin{pmatrix}1&2&3\\3&2&1\end{pmatrix}$$

Obviously, fg ≠ gf.

Example 42(c):

If $f=\begin{pmatrix}1&2&3&4&5\\2&3&4&5&1\end{pmatrix}$ and $g=\begin{pmatrix}1&2&3&4&5\\1&2&4&5&3\end{pmatrix}$ *be two permutations of degree 5, Find fg.*

Solution:

$$fg=\begin{pmatrix}1&2&3&4&5\\2&3&4&5&1\end{pmatrix}\begin{pmatrix}1&2&3&4&5\\1&2&4&5&3\end{pmatrix}$$

$$=\begin{pmatrix}1&2&3&4&5\\2&3&4&5&1\end{pmatrix}\begin{pmatrix}2&3&4&5&1\\2&4&5&3&1\end{pmatrix}$$

$$=\begin{pmatrix}1&2&3&4&5\\2&4&5&3&1\end{pmatrix}$$

Example 43:

If $f=\begin{pmatrix}a&b&c\\b&c&a\end{pmatrix}$ and $I=\begin{pmatrix}a&b&c\\a&b&c\end{pmatrix}$ *be two permutations of degree 3, I being the identity permutation of degree 3, prove that fI = If = f.*

Solution:

$$fI=\begin{pmatrix}a&b&c\\b&c&a\end{pmatrix}\begin{pmatrix}a&b&c\\a&b&c\end{pmatrix}$$

$$=\begin{pmatrix}a&b&c\\b&c&a\end{pmatrix}\begin{pmatrix}b&c&a\\b&c&a\end{pmatrix}$$

$$=\begin{pmatrix}a&b&c\\b&c&a\end{pmatrix}=f$$

and $$If=\begin{pmatrix}a&b&c\\a&b&c\end{pmatrix}\begin{pmatrix}a&b&c\\b&c&a\end{pmatrix}$$

$$= \begin{pmatrix} a & b & c \\ a & c & a \end{pmatrix} = f$$

Hence $f_1 = If = f$.

Example 44:

Show that the multiplicative group G consisting of the three cube roots of unity 1, ω, ω² is isomorphic to the group G' residue classes (mod 3) under addition of residue classes (mod 3.)

Solution:

Let us construct the composition tables of two structures G, G' as given below:

•	1	ω	ω^2
1	1	ω	ω^2
ω	ω	ω^2	1
ω^2	ω^2	1	ω

+(mod3)	[0]	[1]	[2]
[0]	[0]	[1]	[2]
[1]	[1]	[2]	[0]
[2]	[2]	[0]	[1]

From these tables it is evident that if 1, ω, ω^2 are replaced by {0}, {1} {2} respectively in the composition table for G we get the composition table for G'.

This leads to the fact that mapping f of G onto G', defined by f(l) = {0}, f(ω) = {1}, f(ω^2) = {2} is an isomorphism.

$$\text{Also} \quad f(\omega, \omega^2) = f(l) = \{0\} = \{1\} + \{2\}$$
$$= f(\omega) + f(\omega^2)$$

Example 45:

Show that the additive group G = {..., –3, –2, –1, 0, 1, 2, 3, ...} is isomorphic to the additive group

G' = {..., –3m, –2m, –m, 0, m, 2m, 3m, 3m.,} for any given integer m.

Solution:

We define a mapping f by $f : G \to G' : f(a) = ma$,

$$a \in G, ma \in G'$$

we have to show that f is an isomorphism of G onto G'.

We see that f is one-one since two different elements of G have two different f-images in G'. It is also onto since every element of G' is the f-image of an element of G.

Again $\quad f(a + b) = m(a + b)$ [by definition of f]

$= ma + mb$ [by distribution law in I]

$= f(a) + f(b)$

Thus f is composition preserving as well.

Hence f is an isomorphic mapping of G onto G'.

Example 46(a):

Let G be the additive group of real numbers and G', the multiplicative group of all positive real numbers. Show that the following mappings are isomorphic.

(i) $f : G \to G' : f(x) = e^x, x \in G$

(ii) $g : G' \to G : g(x) = \log x, x \in G'$.

Solution:

(i) Since $x \neq y \Rightarrow e^x \neq e^y \ \forall\ x, y \in G$, the mapping f is one-one.

The mapping f is also onto since to every positive real number $x \in G'$, there is a real number $\log_e x \in G$ such that

$$f(\log_e x) = e^{\log ex} = x$$

Moreover $x, y \in G$, we have

$$f(x + y) = e^{x+y} \quad \text{[by definition of f]}$$

$$= e^x \cdot e^y$$

$$= f(x)\, f(y) \quad \text{[by definition of f]}$$

This shows that f preserves compositions in G and G'.

Hence f is an isomorphism of G and G'

(ii) The mapping g is one-one since any two different real positive numbers have two different logarithms.

It is onto since every real number has a unique anti-logarithm as a positive real number, i.e.,

$$a \neq b \Rightarrow \log a \neq \log b \ \forall\ a, b \in G'$$

and $\forall\ x \in G$ we have $x = \log_e e^x$

$= g(e^x)$, where $e^x \in G'$.

Also, $\forall\ x, y \in G'$, we have

$$g(xy) = \log(xy) \quad \text{[by definition of g]}$$

$$= \log x + \log y$$

$$= g(x) + g(y) \quad \text{[by definition of g]}$$

Hence g is an isomorphic mapping of G' onto G.

Example 46(b):

If G be the group {0, 1, 2. 3, 4} under addition modulo five and G' the cyclic group of order five, [a, a^2, a^3, a^4, a^5 = e}, then prove that the mapping

$$f : G \to G' : (n) = a^n \ \forall n \in G \text{ if an isomorphism of } G \text{ onto } G'.$$

Solution:

The mapping f is clearly one-one onto. To see whether f is composition preserving, we set up the composition tables of G and G', keeping in view that a^5 = e and die definition of addition modulo m.

+5	0	1	2	3	4
0	0	1	2	3	4
1	1	2	3	4	0
2	2	3	4	0	1
3	3	4	0	1	2
4	4	0	1	2	3

•	e	a	a^2	a^3	a^4
e	e	a	a^2	a^3	a^4
a	a	a^2	a^3	a^4	e
a^2	a^2	a^3	a^4	e	a
a^3	a^3	a^4	e	a	a^2
a^4	a^4	e	a	a^2	a^3

It is now obvious from the two tables that if we replace 0, 1, 2, 3, 4, by e, a, a^2, a^3, a^3, a^4 in the composition table for G, we get the complete table for G'.

We can verify from these tables that

$$f(m +_5 n) = f(m) f(n) \ \forall m, n \in G.$$

Thus for example, we have

$$f(3) = a^3 \text{ and } f(4) = a^4.$$

Then $\quad f(3) + f(4) - f(2) = a^2$

and $\quad f(3) . f(4) = a^3 . a^4 = a^7$

$$= a^5 . a^2 = a^2$$

Hence $f(3 +_5 4) = f(3) . f(4)$ etc.

The mapping f is therefore isomorphic.

Example 47(a):

Find the regular permutation group isomorphic to the multiplicative group.

$$G = \{1, -1, i, -i\}$$

Solution:

Using Cayley's theorem, we see that G' of the following four permutation :

P_1, P_2, P_3 and P_4.

$$P_1 = \begin{pmatrix} 1 & -1 & i & -i \\ 1.1 & 1.(-1) & 1.i & 1.(-i) \end{pmatrix}$$

$$= \begin{pmatrix} 1 & -1 & i & -i \\ 1 & -1 & i & -i \end{pmatrix} = I$$

$$P_2 = \begin{pmatrix} 1 & -1 & i & -i \\ (-1).1 & (-1).(-1) & (-1).i & (-1).(-i) \end{pmatrix}$$

$$= \begin{pmatrix} 1 & -1 & i & -i \\ -1 & 1 & -i & i \end{pmatrix} = (1- i)\ (i - i),$$

$$P_3 = \begin{pmatrix} 1 & -1 & i & -i \\ i.1 & i.(-1) & i.i & i.(-i) \end{pmatrix}$$

$$= \begin{pmatrix} 1 & -1 & i & -i \\ i & -i & -1 & 1 \end{pmatrix} = (1, -1 - i)$$

$$P_4 = \begin{pmatrix} 1 & -1 & i & -i \\ (-i).1 & (-i).(-1) & (-i).i & (-i).(-i) \end{pmatrix}$$

$$= \begin{pmatrix} 1 & -1 & i & -i \\ -i & i & 1 & -1 \end{pmatrix}$$

$$= (1, -i - 1i)$$

Example 47(b):

*Show that if every element of a group **G** is its own inverse, then **G** is abelian.*

Solution:

Let a and b be any two elements of **G**. Then ab is also an element of **G**.

Therefore, $(ab)^{-1} = ab$ (because it is given that every element in **G** is its own inverse)

Hence, $\quad (ab)^{-1} = ab \Rightarrow b^{-1}\, a^{-1} = ba$

$\Rightarrow$ $ba = ab$ $[\because a^{-1} = a, b^{-1} = b]$

Hence, we have

$$ab = ba \ \forall \ a, b \in G$$

$\therefore$ G is an abelian group.

Example 47(c):

Prove that the set G = (0, 1, 2, 3, 4} is a finite abelian group of order 5 with respect to addition modulo 5.

Solution:

Let us prepare a composition table as given below :

+ (mod 5)	0	1	2	3	4
0	0	1	2	3	4
1	1	2	3	4	0
2	2	3	4	0	1
3	3	4	0	1	2
4	4	0	1	2	3

Closure Property : All the entries in the composition table are elements of the set G. Hence G is closed under addition modulo 5.

Associative Property : Addition modulo 5 is associative always.

Identity : $0 \in G$ is the identity element.

Inverse. It is clear from composition table.

Element	–	0	1	2	3	4
Inverse	–	0	4	3	2	1

Inverse exists for every element of G.

Commutative law : The composition is commutative as the corresponding rows and columns in the table are identical.

The number of elements in G are 5.

Hence (G, + (mod 5)} is a finite abelian group of order 5.

Example 47(d):

Prove that the set G = {1, 2, 3, 4, 5, 6} is a finite abelian group of order 5 with respect to multiplication modulo 7.

Solution:

Let us prepare the following composition table :

$\times_7$	1	2	3	4	5	6
1	1	2	3	4	5	6
2	2	4	6	1	3	5
3	3	6	2	5	1	4
4	4	1	5	2	6	3
5	5	3	1	6	4	2
6	6	5	4	3	2	1

Closure Property : All the entries in the table are elements of **G**. Therefore **G** is closed with respect to multiplication modulo 7.

Associative Property : Multiplication modulo 7 is associative always.

Identity : Since first row of the table is identical to the row of elements of **G** in the horizontal border, the element to the left of first row in vertical border is identity element, i.e., 1 is identity element in G with respect to multiplication modulo 7.

Inverse : From the table it is obvious that inverses of 1, 2, 3, 4, 5, 6 are 1, 4, 5, 2, 3 and 6 respectively. Hence inverse of each element in G exists.

Commutative Property : The composition is commutative because the elements equidistant from principal diagonal are equal each to each.

The set **G** has 6 elements. Hence $(G, \times_7)$ is a finite abelian group of order 6.

Example 48:

Show that a finite set G with an associative binary operation is a group if the right and left cancellation laws hold in G. Show further that validity of the cancellation laws does not characterize infinite groups but characterize finite groups only.

Solution:

Let $G = \{a_1, a_2, a_3 ..., a_n\}$ where $a_1, a_2, ... a_n$ are distinct. Let us consider an element a of **G** then $a_1a, a_2a, a_2a, ... a_n a \in G$ the set **G** being closed under the binary operation and these elements are all distinct, because, otherwise $a_ia = a_ja$, for $a_i, a_j \in G$ $(i \neq j)$ for $a_i = a_j$ by the right cancellation law.

Thus $a_1a, a_2a, a_3a, \ldots, a_n a$ are nothing but some permutation of elements $a_1, a_2, a_3, ... a_n$ of **G**.

Therefore, if b is an element of **G**, then it must be one of the elements $a_1a, a_2a, ... a_na$.

Hence $a_ia = b$, where $a, b, a_i \in G$.

Thus we arrive at the conclusion that the equation ya = b(a, b $\in$ G) has a unique solution in **G**. Similarly by forming the products $aa_1, aa_2, aa_3, \ldots aa_n$ and by using the left cancellation law,

We can show that the equation ax = b (a, b $\in$ **G**) has a solution in **G**.

Hence **G** is a finite group of order n.

However, an infinite set will not necessarily form a group even if the binary composition in a set is associative and both the cancellation laws hold good in it. The following example will illustrate this fact.

The set **N** of all natural numbers 1, 2, 3, 4, ... is not a group for multiplication even though N is closed under multiplication, multiplication is associative and both the cancellation laws hold good for multiplication in **N**.

Example 49(a):

Show that the set of all integers .... –4, –3, –2, –1, 0, 1, 2, 3, 4 .... is an infinite abelian group with respect to the operation of addition of integers.

Solution:

Let us test all the group axioms for abelian group.

(G_1) Closure axiom : We know that the sum of any two integers is also an integer, i.e., for all a, b $\in$ **I**, a + b $\in$ **I**. Thus I is *closed* with respect to addition.

(G_2) Associativity : Since the addition of integers is *associative*, the associative axiom is satisfied, i.e., for a, b, c $\in$ **I**.

$$a + (b + c) = (a + b) + c$$

(G_3) Existence of Identity : We know that 0 is the additive identity and 0 $\subset$ **I**, **i.e.,**

$$0 + a = a = a + 0 \ \forall \ a \in \mathbf{I}$$

Hence additive identity exists.

(G_4) Existence of Inverse : If a $\in$ **I**, then –a $\in$ **I**. Also,

$$(-a) + a = 0 = a + (-a).$$

The every integer possesses additive inverse.

Therefore I is a *group* with respect to addition.

Since addition of integers is a commutative operation, therefore a + b = b + a $\forall$ a, $\in$ **I**.

Hence (**I**, +) is *an abelian group*. Also, I contains an infinite number of elements. Therefore (**I**, +) is an abelian group of infinite order.

Example 49(b):

Prove that the set of cube roots of unity is an abelian finite group with respect to multiplication.

Solution:

The set of cube roots of unity is, **G** = $\{(1, \omega\omega^2)\}$. Let us form the composition table as given below:

•	1	ω	ω^2
1	1	ω	ω^2
ω	ω	ω^2	$\omega^3 = 1$
ω^2	ω^2	$\omega^3 = 1$	$\omega^4 = \omega$

(G_1) **Closure axiom :** Since each element obtained in the table is a unique element of the given set **G**, multiplication is a binary operation. Thus the closure axiom is satisfied.

(G_2) **Associative axiom :** The elements of G are all complex numbers and we known that multiplication of complex numbers is always associative. Hence associative axiom is also satisfied

(G_3) **Identity axiom :** Since row 1 of the table is identical with the top border row of elements of the set, 1 (the element of the extreme left of this row) is the identity element in G.

(G_4) **Inverse axiom :** The inverse of 1 ω, ω^2 are 1, ω^2 and ω respectively.

(G_5) **Commutative axiom :** Multiplication is commutative in **G** because the elements equi-distant with the main diagonal are equal each to each.

The number of elements in **G** is 3. Hence (**G**,.) is a finite group of order 3.

Example 50(a):

Show that the set of all even integers (including zero) with additive property is an abelian group.

Solution:

The set of all even integers (including zero) is

$$I = \{0, \pm 2, \pm 4, \pm 6....)$$

Now we will discuss the group axioms one by one :

(G_1) The sum of two even integers is always an even integer, therefore *closure axiom* is satisfied.

(G_2) The addition is associative for even integers, *hence associative axiom* is satisfied.

(G_3) **0** $\in$ **I**, which is an additive identity in **I**, hence *identity axiom* is satisfied.

(G_4) Inverse of an even integer a is the even integer –a in the set, so *axiom of inverse* is satisfied.

(G_5) Commutative law is also satisfied for addition of even integers. Hence the set forms an abelian group.

Example 50(b):

Show that the set of all non-zero rational numbers with respect to binary operation multiplication is a group.

Solution:

Let the given set be denoted by $\mathbf{Q_0}$. Then by group axioms, we have

(G_1) We know that the product of two non-zero rational numbers is also a non-zero rational number. Therefore, $\mathbf{Q_0}$ is closed with respect to *multiplication.*

Hence, closure axiom is satisfied.

(G_2) We know for rational numbers.

$$(a \,.\, b) \,.\, c = a \,.\, (b \,.\, c) \text{ for all } a, b, c \in \mathbf{Q_0}$$

Hence, *associative axiom* is satisfied.

(G_3) Since, 1 the multiplicative identity is a rational number hence *identity axiom* is satisfied.

(G_4) If a $\in \mathbf{Q_0}$, then obviously, 1/a $\in \mathbf{Q_0}$. Also

$$1/a \,.\, a = 1 = a \,.\, 1/a$$

so that 1/a is the multiplicative inverse of a. Thus *inverse axiom* is also satisfied.

Hence $\mathbf{Q_0}$ is a group with respect to multiplication.

Example 50(c):

Show that C, the set of all non-zero complex numbers is a multiplicative group.

Solution:

Let $\mathbf{C} = \{z : z = x + iy,\ x, y \in \mathbf{R}\}$

Hence **R** is the set of all real numbers and $i = \sqrt{(-1)}$.

(G_1) **Closure axiom :** If a + ib $\in$ **C** and c + id $\in$ **C**, then by definition of multiplication of complex numbers

$$(a + ib)\,(c + id) = (ac - bd) + i\,(ad + bc) \in \mathbf{C}.$$

since $ac - bd, ad + bc \in \mathbf{R}$, for $a, b, c, d \in \mathbf{R}$.

Therefore, **C** is closed under multiplication.

(G_2) **Associative Axiom:**

$(a + ib) \{(c + id) . (e + if)\} = (ace - adf - bcf - bdc) + i(acf + ade + bce - bdf)$

$$= \{(a + ib) . (c + id)) . (e + if)$$

for $a, b, c, d \in \mathbf{R}$.

(G_3) **Identity Axiom :** $e = 1 (= 1 + i. 0)$ is the identity in **C**.

(G_4) **Inverse Axiom :** Let $(a + i b) (\neq 0) \in \mathbf{C}$, then

$$(a+ib)^{-1} = \frac{1}{a+ib} = \frac{a-ib}{a^2+b^2}$$

$$= \left(\frac{a}{a^2+b^2}\right) + i\left(-\frac{b}{a^2+b^2}\right)$$

$$= m + i n \in \mathbf{C}, \text{where } m = \left(\frac{a}{a^2+b^2}\right), n = -\frac{b}{a^2+b^2} \in R$$

Hence **C** is a multiplicative group.

Example 51(a):

If S be the set of all real numbers of the form $(m + n\sqrt{2})$ where m, $n \in Q$, set of rational numbers, prove that S is a multiplicative or additive group, m, n not vanishing simultaneously.

Solution:

Let us first prove the question for multiplication as composition.

(G_1) Suppose $m_1 + n_1\sqrt{2}, m_2 + n_2\sqrt{2} \in S$, so that $m_1, n_1, m_2, n_2 \in R$.

Now,

$(m_1 + n_1\sqrt{2}).(m_2 + n_2\sqrt{2}) = (m_1m_2 + 2n_1n_2) + (m_1n_2 + m_2n_1)\sqrt{2}$

Which belongs to S because $(m_1m_2 + 2n_1n_2), (m_1n_2 + m_2n_1) \in \mathbf{Q}$.

Hence *closure axiom* is satisfied.

(G_2) The elements of S are all real-numbers and the multiplication of real numbers is associative.

Hence *associative axiom* is satisfied.

(G_3) $1 = (1 + 0. \sqrt{2}) \in S$ is an identity element to multiplication.

(G_4) If $(m + n\sqrt{2}) \in S$ then

$$(m+n\sqrt{(2)})^{-1} = \frac{1}{m+n\sqrt{(2)}} = \frac{m-n\sqrt{2}}{m^2-2n^2}$$

$$= \left(\frac{m}{m^2-2n^2}\right) + \left(\frac{-n}{m^2-2n^2}\right)\sqrt{2}$$

$$= \alpha + \beta\sqrt{2} \in \mathbf{S}$$

Where $\alpha = \frac{m}{m^2-2n^2}, \beta = \frac{-n}{m^2-2n^2} \in \mathbf{Q} \ \forall \ \mathbf{m, n} \in \mathbf{Q}.$

Hence the *inverse axiom* is satisfied.

Therefore (S,.) is a group. Similarly, we can slow that (S, +) is also a group.

Example 51(b):

Show that the set of all n × n non-singular matrices having their elements as rational (real or complex) number is an infinite non-abelian group with respect to matrix multiplication.

Solution:

Let M be the set of all n × n non-singular matrices with their elements as rational numbers.

Let **A** ∈ **M** ⇒ A is a square matrix of order n, whose elements are rational numbers and $|A| \neq 0$, because it is given that members of **M** are non-singular.

(G_1) Closure Property : If A, B ∈ M then AB will also be a matrix of order n × n, the elements of A will all be rational numbers, also $|AB| = |A|\,|B| \neq 0$ as $|A| \neq 0$ and $|B| \neq 0$.

Thus AB ∈ M for A, B ∈ M. Therefore M is closed under matrix multiplication.

(G_2) Associativity : Multiplication of matrices is associative always.

(G_3) Existence of Identity : If I be the unit matrix of the type n × n, then the elements of I are all rational numbers. Also $|I| = 1$, i.e., $\neq 0$.

Therefore I ∈ M and also I A = A I ∀ A ∈ M.

Therefore I is identity element.

(G_4) Existence of Inverse : We known that inverse exist for all non-singular matrices. Therefore, if A ∈ M, there exists a non-singular matrix

$A^{-1} = \dfrac{\text{Adj}A}{|A|}$ with elements as rational numbers such that $\mathbf{A^{-1}\,A = I =}$

$\mathbf{AA^{-1}}$.

Matrix multiplication, is not in general commutative. Number of elements in **M** is infinite.

Therefore **M** is an infinite non-abelian group with respect to matrix multiplication.

Example 52:

Prove that the set of matrices

$$A_\alpha = \begin{bmatrix} \text{Cos}\,\alpha - \text{Sin}\,\alpha \\ \text{Sin}\,\alpha - \text{Cos}\,\alpha \end{bmatrix}$$

where α *is a real number, forms a group under multiplication.*

Solution:

(G_1) Closure Property : Let $A_\alpha, A_\beta \in G$. Then

$$A_\alpha A_\beta = \begin{bmatrix} \text{Cos}\,\alpha & -\text{Sin}\,\alpha \\ \text{Sin}\,\alpha & \text{Cos}\,\alpha \end{bmatrix} \begin{bmatrix} \text{Cos}\,\beta & -\text{Sin}\,\beta \\ \text{Sin}\,\beta & \text{Cos}\,\beta \end{bmatrix}$$

$$= \begin{bmatrix} \text{Cos}\,(\alpha+\beta) & -\text{Sin}\,(\alpha+\beta) \\ \text{Sin}\,(\alpha+\beta) & \text{Cos}\,(\alpha+\beta) \end{bmatrix}$$

$$= A_{\alpha+\beta} \in G$$

(as $\alpha + \beta$ is a real number whenever α and β are real).

(G_2) Associativity. Let A_α, A_β and A_γ belong to G. Then

$$(A_\alpha \,.\, A_\beta)A_\gamma = A_{\alpha+\beta} \,.\, A_\gamma = A_{(\alpha+\beta)+\gamma}$$
$$= A_{\alpha+(\beta+\gamma)} = A_\alpha \,.\, (A_\beta \,.\, A_\gamma)$$

(G_3) Identity axiom : Since 0 is a real number, therefore

$$A_0 = \begin{bmatrix} \text{Cos}\,0 & -\text{Sin}\,0 \\ \text{Sin}\,0 & \text{Cos}\,0 \end{bmatrix} = \begin{bmatrix} 1 & 0 \\ 0 & 1 \end{bmatrix} \in G$$

Now, $\quad \mathbf{A_\alpha \,.\, A_0 = A_{\alpha+0} = A_\alpha}$

Thus, $\quad \mathbf{A_0 = I_2}$ is the identity element.

(G_4) Inverse Axiom : Let $\mathbf{A_\alpha \in G}$, α is a real number. Then $\mathbf{A_{-\alpha} \in G}$, since $-\alpha$ is also a real number, and we have

$$A_\alpha \cdot A_{-\alpha} = A_{(\alpha+(-\alpha))} = A_0 = A_{-\alpha} \cdot A_\alpha$$

Hence $A_{-\alpha}$ is the inverse of A_α.

Thus G is a group under matrix multiplication.

Example 53:

Prove that the set Q of all rational numbers other then 1 with the operation defined by a o b = a + b – ab constitutes an abelian group.

Solution:

(G_1) Closure Property : Let a, b ∈ **Q** so that both a and b are rational numbers other than 1.

∴ a o b = a + b – ab is also a rational number other than 1 because if it is equal to 1,

i.e., if $a + b - ab = 1$ then $a + b - ab - 1 = 0$

or $(a - 1)(b - 1) = 0$.

This would mean that as = 1 or b = 1 which is not true. Hence aob ∈ **Q**, i.e., the closure property is satisfied.

(G_2) Associative Law : Let a, b, c ∈ **Q**, then

$$(a \circ b) \circ c = (a + b - ab) \circ c$$
$$= (a + b - ab) + c - (a + b - ab)c$$
$$= a + b + c - ab - ac - bc + abc.$$

Also, $$a \circ (b \circ c) = a \circ (b + c - bc)$$
$$= a + (b + c - bc) - a(b + c - bc)$$
$$= a + b + c - ab - ac - bc + abc.$$

Thus (a o b) o c s = a o (b o c) and hence associative law is satisfied.

(G_3) Existence of Identity : If e be the identity then

$$a \circ e = a \Rightarrow a + e - ae = a$$

$$\Rightarrow \quad e(1 - a) = 0 \quad e = 0 \text{ as } a \neq 1$$

and in the case $a \circ e = a \circ 0 = a + 0 - a.\,0 = a$

Hence 0 is the identity.

(G_4) Existence of Inverse :

$$a \circ b = a + b - ab = 0 \text{ the identity.}$$

$\Rightarrow \quad \dfrac{a}{a-1} = b$ is the inverse of a

Also, since $a - 1 \neq 0$

$\therefore$ a/a – 1 is a rational number other than 1 and as such belongs to **Q**,

Therefore inverse exists and belongs to the set. •

(G_5) Commutative Law :

$$a \text{ o } b = a + b - ab = b + a - ba = b \text{ o } a$$

Hence commutative.

Therefore, **Q** is an abelian group.

EXERCISES

1. If H be a sub-group of G and N be a normal sub-group of G, then prove that $H \cap N$ is a normal subgroup of H.
2. Show that the additive group of all integers is isomorphic to the additive group of all even integers.
3. Show that the group G = [{1, –1}] is isomorphic to the group

   $G' = [\{f_1, f_2\}, 0]$

   where $f_1 : R \to R : f_1(x) = x,\ f_2 : R \to R : f_2(x) = -x$.
4. Show that the set of all odd integers with addition as operation is not a group.
5. Prove that the four matrices :

   $$\begin{bmatrix} 1 & 0 \\ 0 & 1 \end{bmatrix}, \begin{bmatrix} -1 & 0 \\ 0 & 1 \end{bmatrix}, \begin{bmatrix} 1 & 0 \\ 0 & -1 \end{bmatrix}, \begin{bmatrix} -1 & 0 \\ 0 & -1 \end{bmatrix}$$

   form a multiplication group.
6. Prove that the set M of complex numbers z with condition $|z| = 1$ forms a group with respect to the operation of ordinary multiplication.
7. Show that the nth roots of unity form a finite abelian group of order n with respect to multiplication.
8. Prove that for every element a is a group G, $a^2 = e$, where e is the identity, then G is an abelian group.
9. Show that if a, b are any two elements of a group G, then $(ab)^2 = a^2b^2$, if an only if G is abelian.
10. Show that the set of all numbers $\cos\theta + i \sin\theta$ forms an infinite abelian group with respect to ordinary multiplication; where θ runs over all rational numbers.

11. Show that the residue classes modulo 5 form a group with respect to addition.

12. Show that the residue classes modulo 6 with respect to multiplication do not form a group.

13. Show that the set of residue classes modulo m is an abelian group of order m with respect to addition of residue classes.

14. Show that the set $\{0, 1, 2, 3\}$ is a finite abelian group of order 4 under addition modulo 4 as composition.

15. Write down all permutations on three symbols a, b, c. Which of these permutations are even?

16. If a cyclic sub-group N of G is normal in G, then prove that every sub-group of N is normal in G.

17. Prove that the multiplicative group (1, –1, i, –i) is isomorphic to the permutation group {I, (a b c d), (ac) (bd), (a d c b)} on four symbols a, b, c, d.

18. Prove that the set A_3 consisting of three permutations I, (a b c), (acb) on three symbols a, b, c form a finite abelian group with respect to the permutation multiplication. Also prepare a composition table.

19. Show that integral multiples of 5 form a sub - group of the additive group of all integers including 0.

20. Show that the set of all odd integers with the binary operation o defined as aob = a – b (a, b ∈ I) is not a group.

21. Show that the set

    $$G = \{ ...., -4m, -3m, -2m, -m, 0, m, 2m\ 3m, 4m, ...\}$$

    of multiples of integers by a fixed integer m is a group with respect to addition.

22. Let G be the multiplicative group of all positive real numbers and R the additive group of all real numbers. Is G a sub-group of R?

23. Prove that there is a one one correspondence between the set of left cosets of H and the set of right cosets of H.

24. If H be a sub-group of a group G, and a ∈ G, then show that

    $(Ha)^{-1} = a^{-1} H$.

26. Prove that the set of four transformations actions f_1, f_2, f_3, f_4, defined by $f_1(z) = z$, $f_2(z) = -z$, $f_3(z) = 1/z$, $f_4(z) = -1/z$ forms a finite abelian group with respect to the composite composition.

27. G = {[0], [1], [2], [3]} is the additive group of residue classes modulo 4. Find the order or each element.

28. Prove that a group G is abelian if every element of G (except the identity) is of order 2.

28. If a group G has four elements, show that it must be abelian.

30. Show that the fifth roots of unity form a cyclic group with respect to multiplication, the generator being $e^{2\pi/5}$.

31. By using Lagrange's theorem for finite groups, prove that {0, 1, 2, 3} is not a sub-group of (Z_9, $+_9$).

32. How many elements of the cyclic group of order ∞ can be used as generators of the group?

33. Show that {1, –1, i, –i} is a cyclic group under multiplication. Also find its generator.

34. Verify that the totality of all positive rationals form a group under the composition defined by aob = ab/3

3

ALGEBRAIC SYSTEMS

INTRODUCTION OF ALGEBRAIC SYSTEMS

An *algebraic system* is a mathematical system consisting of a set called the *domain* and one or more operations on the domain. If V is the domain and * 1, .., *n are the operations, [V; * 1, ..., n] is used to denote the mathematical system. If the context is clear, this notation is abbreviated to V.

Example 1:

Consider the set of two by real matrices, $M_{2 \times 2}(\mathbf{R})$, with the operation of matrix multiplication. In this context, can be interpreted as saying that if $AB = BA$, $(AB)^2 = A^2B^2$. One pair of matrices that this theorem applies to is

$$\begin{bmatrix} 2 & 1 \\ 1 & 2 \end{bmatrix} \text{ and } \begin{bmatrix} 3 & -4 \\ -4 & 3 \end{bmatrix}$$

In ohter words, the algebraic structures and the axioms which define them have a certain naturality about them. They come from the experience of looking at many examples, and are rich in meaningful results. Not just sitting down and giving a few axioms at random can lead us to results of wide interest and use. A particular system is chosen for study because its numerous examples from the set of numbers, matrices, vectors and many other mathematical objects appear again and again. Two seemingly different mathematical objects. Become analogous when considered as a special case of some algebraic system. We want to be acquainted with the basic concept and properties of an algebraic structure. It is here, in the study of an algebraic structure, that the students experience for the first time the generalization of the fundamental concepts of classical mathematics, with precision and logical reasoning.

These functions can of course be studied individually, but this doesn't carry us very far. It is desirable to consider the sets Q(X, R) and Q(X, C) of all such functions as mathematical systems with a high level of internal organization, and this program compels us to give serious attention to their structural features. It is at this point that algebra enters the picture; for modern algebra is essentially the result of crystallizing into abstract form, and studying for their own sake, a few simple patterns of structure which underlie many diverse parts of mathematics. We defined what is meant by a linear space and an algebra, but we did not develop the theory of these systems to any appreciable degree. We used them only descriptively, as a convenient means of calling attention to the fact that the points in the spaces R" and C^n can be added and multiplied by numbers, and those in Q(X, R) and Q(X, C) can be multiplied together as well. Our work in the rest of this book requires a deeper understanding of these systems and several others, and the purpose of this chapter is to provide a concise but reasonably complete exposition of this necessary background material.

The algebraic systems we discuss below—groups, rings, linear spaces, and algebras—have been the subject of many books and innumerable research articles. We clearly can do little more than explain what each system is, mention several outstanding examples, and develop the theory to the limited extent required by our later work. If the reader finds it desirable to amplify our abbreviated treatment by consulting additional sources, we suggest McCoy [31] and Halmos [17].

Example 2:

(a) Let B* be the set of all finite strings of 0's and 1's including the null (or empty) string λ. An algebraic system is obtained by adding the operation of *concatenation*. The concatenation of two strings is simply the linking of the two strings together in the order indicated. The concatenation of a with b is often denoted (a) (b) with no operation symbol. For example (01101)(101) = 01101101 and (λ)(100) = 100. Note that concatenation is an associative operation and that λ is the identity for concatenation.

(b) Let M be any non-empty set and let * be any operation on M that is associative and for which an identity exists in M.

Our second example might seem strange, but we include it to illustrate a point. The algebraic system B* is a special case of [M; *]. Most of us are much more comfortable with B* than with M. No doubt the reason is that the elements in B* are more concrete. We know what they look like and exactly how they are combined. The description of M is so vague that we don't even know what the elements are, much less how they are combined. Why would anyone want to study M? The reason is related to this question: What theorems are of interest in an algebraic system? Answering this question is one of our main objectives in this chapter. Certain properties of algebraic systems are called Algebraic properties, and any theorem that says something about the algebraic properties of a system would be of interest. The ability to identify what is algebraic and what isn't is one of the skills that you should learn from this chapter.

Now, back to the question of why we study M. Our answer is to illustrate the usefulness of M with a theorem about M.

Theorem:

*If a, b are elements of M and a * b = b * a, then (a * b) * (a * b) = (a * a) * (b * b).*

Proof:

(a * b) * (a * b)	
= a * (b * (a * b)	Why?
= a * ((b * a) * b)	Why?
= a * ((a * b(* b)	Why?
= a * (a * (b * b))	Why?
= (a * a) * (b * b)	Why?

The power of this theorem is that it can be applied to any algebraic system the M describes. Since B* is one such system, we can apply to any two strings that commute—for example, 01 and 0101. Although a special case of this theorem could have been proven for B*, it would not have been any easier to prove, and it would not have given us any insight into other special cases of M.

BINARY OPERATION ON A SET

Let G be a non-empty set. Then $G \times G = \{(a, b) : a \in G, b \in G\}$. If $f : G \times G \rightarrow G$ then f is said to be a binary operation on the set G. The image of the ordered pair (a, b) under the function f is denoted by *afb*. Generally the symbols '+', '×', '.', '0', '*', etc. are used to denote binary operations on a set.

Thus an operation which combines two elements of a set to give another element of the same set is called a *'binary operation'* or a *'binary composition'*.

For example addition is a binary operation on the set N of natural numbers because the sum of two natural numbers is always a natural number. If we take 3 and 6 the two elements belonging to N then $3 + 6 = 9 \in N$. Subtraction is not a binary operation on N because $3, 5 \in N$ and

$3 - 5 = -2 \notin N$. No doubt subtraction is a binary operation on the set of integers I.

If R be the set of all real numbers and a, b be two elements of R then $a + b \in R$ and $a.b \in R$.

General notation 'o' or '*' is used to represent a binary operation. Thus, if 'o' stands for addition then a o b $\Rightarrow$ a + b and if it stands for multiplication aob $\Rightarrow$ a.b etc.

From the above discussion it is concluded that a *binary operation on a set is a rule which assigns to any two elements of the set a unique element in the set.*

Hence if 'o' is a binary operation in the set, the set is said to be closed with respect to the operation o, *i.e.*, if for a, b R $\Rightarrow$ aob $\in$ R then R is closed under operation o.

The operations of algebraic addition, subtraction, multiplication and division can now be looked upon in this reference over the set of numbers. Evidently aob = a ÷ b or a/b is a binary operation over the set of non-zero rational numbers. Addition and multiplication are binary operations on the sets of natural numbers, integers, rational numbers and real numbers. The most convenient notation to denote a binary composition is that of multiplication. In this notation if a, b $\in$ G then ab represents the element obtained on multiplying a and b. Thus a b $\in$ G $\forall$ a, b $\in$ G if the binary operation is G has been denoted multiplicatively.

SOME GENERAL PROPERTIES OF GROUPS

In this section, we will present some of the most basic theorems of group theory. Keep in mind that each of these theorems tells us something about every group. We will illustrate this point at the close of the section.

Theorem 1:

The inverse of any element of a group is unique.

Proof:

The same problem is encountered here as in the previous theorem. We will leave it to the reader to rephrase this theorem. The proof is also left to the reader to write out in detail. Here is a hint: If b and c are both inverse of a, then you can prove that b = c. If you have difficulty with this proof, note that we have already proven it i a concrete setting.

The significance of Theorem 1 is that we can refer to the inverse of an element without ambiguity. The notation for the inverse of a is usually a^{-1} (note the exception below).

Theorem 2:

The identity of a group is unique.

Proof:

One difficulty that students often encounter is how to get started in proving a theorem like this. The difficulty is certainly not in the theorem's complexity. Before actually starting the proof, we rephrase the theorem so that he implication it states is clear.

Theorem 3(a):

*(Rephrased): If G = [G; *] is a group and e is an identity of G, then no other element of G is an identity of G.*

Proof:

[Indirect]: Suppose that $f \in G$, $f \neq e$, and f is an identity of G. We will show that f = e, a contradiction, which completes the proof.

$$f = f * e \text{ Since e is an identity}$$
$$= e \text{ Since f is an identity. \#}$$

Example:

(a) e^{-1} is the inverse of the identity e, which always is e.

(b) $(a^{-1})^{-1}$ is the inverse of a^{-1}, which is always equal.

(c) $(x * y * z)^{-1}$ is the inverse of $(x * y * z)$.

(d) In a concrete group with an operation that is addition or is similar to addition, the inverse of a is usually written –a. For example, the inverse of x – 3 in the group [**Z**; +] is written – (x – 3). In the group of 2 by 2 matrices over the real numbers, the inverse of

$$\begin{bmatrix} 4 & 1 \\ 1 & -3 \end{bmatrix} \text{ is written } \begin{bmatrix} 4 & 1 \\ 1 & -3 \end{bmatrix}$$

Z_n THE INTEGERS MODULE n

In this section we introduce a collection of concrete groups, one for each positive integer, that will provide us with a wealth of examples and applications.

The Division Property for integers: If $m, n \in Z$, $n > 0$, then there exist two unique integers, q (quotient) and r (remainder), such that $m = nq + r$ and $0 \leq r < n$.

Note: The division property says that if m is divided by n, you will obtain a quotient and a remainder, where the remainder is less than n. This is a tact that most elementary school students learn when they are introduced to long division. In performing the division 1986 ÷ 97, you obtain a quotient of 20 and a remainder of 46; i.e., 1986 = (97) (20) + 46.

If two numbers, a and b, share the same remainder after dividing by n, we say that they are *congruent module n*, denoted a = b (mod n). For example, 13 = 38 (mod 4) because 13 = (5) (2) + 3 and 38 = (5) (7) + 3.

Modular Arithmetic: If n is a positive integer, we define the operations of addition n (× n) as follows. If, a, b ∈ **Z**.

a + n b = remainder after a + b is divided by n, and

a × n b = remainder after a × b is divided by n.

Notes:

(a) The result of doing arithmetic module n is always an integer between 0 and n – 1 (Why?). This observation implies that {0, 1, ... n – 1} is closed under module n arithmetic.

(b) It is always true that a + nb = (a + b) (mod n) and a × nb = (a × b) (mod n). For example, $4 +_7 5 = 2 = 9$ (mod 7) and $4 \times_7 5 = 6 = 20$ (mod 7).

(c) We will use the notation $\mathbf{Z}_n$ to denote the set {0, 1, 2, ... n – 1}.

PROPERTIES OF MODULAR ARITHMETIC ON Z_N

Addition module n is always commutative an associative; 0 is the identity for + n and every element of $\mathbf{Z}_n$ has an additive inverse.

Multiplication module n is always commutative and associative, and 1 is the identity for ×

Multiplication module n is distributive (both right and left) over addition module n.

Theorem:

If a ∈ Zn, a ≠ 0, then –a = n – a.

Proof:

a + n(n – a) = (a + (n – a) (mod n) = n (mod n). The only, element of $\mathbf{Z}_n$ that is congruent to n is 0. #

Note: The algebraic properties of + n and × n on Z_n are identical to the properties of addition and multiplication on **Z**.

The group $\mathbf{Z}_n$: For each n ≥ 1, $[Z_n; +_n]$ is a group. Henceforth, we will use the notation Zn when referring to this group. Figure 11.4.1. contains the tables for Z_1 through Z_5.

Example:

(a) *We are all somewhat familiar with* Z_{12} *since the hours of the day are counted using this group, except for the fact that 12 is used in place of 0. If someone started a four-hour trip at 10 o'clock, she would arrive at* $10 -_{12} 4 = 2$ *o'clock. If a satellite orbits the earth every two hours and starts its first orbit at 5 o' clock, it would end its first orbit at 5 + 122 = 7 o'clock. Its seventh orbit would end at at 5 + 12 7 (2) = 7 o'clock.*

$+_1$	0
0	0

$+_2$	0	1
0	0	1
1	1	0

$+_3$	0	1	2
0	0	1	2
1	1	2	0
2	2	0	1

$+_4$	0	1	2	3
0	0	1	2	3
1	1	2	3	0
2	2	3	0	1
3	3	0	1	0

$+_5$	0	1	2	3	4
0	0	1	2	3	4
1	1	2	3	4	0
2	2	3	4	5	0
3	3	4	0	1	2
4	4	0	1	2	3

(b) Virtually all computers represent unsigned integers in binary form with a fixed number of digits. A very small computer might reserve seven bits to store the value of an integer. There are only 2^7 different values that can bo stored in seven bits. Since the smallest value is 0, represented as 0000000, the maximum value will be $2^7 - 1 = 127$, represented as 11111111. When a command is given to add two integer values, and the two values have a sum of 128 or more, overflow occurs. For example, if we try to add 56 and 95, the sum is an eight-digit binary integer 10010111. One common procedure is to retain the seven lowest-ordered digits. The result of adding 56 and 95 could be $0010111_{two} = 23 = (56 + 95)$ (*mod* 128). Integer arithmetic with this computer would actually be module 128 arithmetic.

OPERATIONS

One of the first mathematical skills that we all learn is how to add a pair of positive integers. A young child soon recognizes that something is wrong if a sum has two values, particularly if his or her sum is different from the teacher's. In addition, it is unlikely that a child would consider assigning a non-positive value to the sum of two positive integers. In other words, at an early age we probably know that the sum of two positive integers is unique and belongs to the set of positive integers. This is what characterizes all operations on a set.

Definition: Binary Operation. *Let S be a non-empty set. A binary operation on s is a rule that assigns to each ordered pair of elements of S a unique element of S. In other words, a binary operation is a function from S × S into S.*

Example:

Union and intersection are both binary operations on the power set of any universe. Addition and multiplication are binary operators on the natural numbers. Addition and multiplication are binary operations on the set of 2 by 2 real matrices, $M_{2\times 2}(\mathbf{R})$. Division is a binary operation on some sets of numbers, like the positive reals. But on the integers ($1 \div 2 \notin \mathbf{Z}$) and even on the reals ($1 \div 0$ is not defined), division is not a binary operations.

Note:

(a) We stress that the image of each ordered pair must be in S. This requirement disqualifies subtraction on the natural numbers from consideration as a binary operation, since 1 – 2 is not a natural number. Subtraction is a binary operation on the integers.

(b) *On Notation.* Despite the fact that a binary operation is a function, symbols, not letters, are used to name them. The most commonly used symbol for a binary operation is an asterisk, *. We will also use $ when a second symbol is needed.

(c) If * is a binary operation on S and a, b ∈ S, there are three common ways of denoting the image of the pair (a,b). They are:

*ab	a * b	Ab *
Prefix Form	Infix Form	Postfix Form

We are all familiar with infix form. For example, 2 + 3 is how everyone is taught to write the sum of 2 an 3. But notice how 2 = 3 was just described in the previous sentence! The word *sum* preceded 2 and 3. Orally, prefix form is quite natural to us. The prefix and postfix forms are superior to infix form in some respects. We saw that algebraic expressions with more than one operation didn't need parentheses if they were in prefix or postfix form. Due to our familiarity with infix form, we will use it throughout the remainder of this book.

Some operations, such as negation of numbers and complementation of sets, are not binary, but binary operators.

Definition: Unary Operation. *Let S be a non-empty set. A unary operator on S is a rule that assigns to each element of S a unique element of S. In other words a unary operator is a function from S into S.*

SOME GENERAL PROPERTIES OF GROUPS

In this section, we will present some of the most basic theorems of group theory. Keep in mind that each of these theorems tells us something about every group. We will illustrate this point at the close of the section.

Theorem 1:

The inverse of any element of a group is unique.

Proof:

The same problem is encountered here as in the previous theorem. We will leave it to the reader to rephrase this theorem. The proof is also left to the reader to write out in detail. Here is a hint: If b and c are both inverse of a, then you can prove that b = c. If you have difficulty with this proof, note that we have already proven it i a concrete.

The significance is that we can refer to the inverse of an element without ambiguity. The notation for the inverse of a is usually a^{-1} (note the exception below).

Theorem 2:

The identity of a group is unique.

Proof:

One difficulty that students often encounter is how to get started in proving a theorem like this. The difficulty is certainly not in the theorem's complexity. Before actually starting the proof, we rephrase the theorem so that he implication it states is clear.

Theorem 3:

*(Rephrased): If G = [G, *] is a group and e is an identity of G, then no other element of G is an identity of G.*

Proof:

[Indirect]: Suppose that f ∈ G, f ≠ e, and f is an identity of G. We will show that f = e, a contradiction, which completes the proof.

$f = f * e$ Since e is an identity

$= e$ Since f is an identity. #

Example:

(a) e^{-1} is the inverse of the identity e, which always is e.

(b) $(a^{-1})^{-1}$ is the inverse of a^{-1}, which is always equal.

(c) $(x * y * z)^{-1}$ is the inverse of $(x * y * z)$.

(d) In a concrete group with an operation that is addition or is similar to addition, the inverse of a is usually written –a. For example, the inverse of x – 3 in the group [Z; +] is written – (x – 3). In the group of 2 by 2 matrices over the real numbers, the inverse of

$$\begin{bmatrix} 4 & 1 \\ 1 & -3 \end{bmatrix} \text{ is written } \begin{bmatrix} 4 & 1 \\ 1 & -3 \end{bmatrix}.$$

THE INTEGERS MODULE n

In this section we introduce a collection of concrete groups, one for each positive integer, that will provide us with a wealth of examples and applications.

The Division Property for integers. If $m, n \in Z$, $n > 0$, then there exist two unique integers, q (quotient) and r (remainder), such that $m = nq + r$ and $0 \leq r < n$.

Note:

The division property says that if m is divided by n, you will obtain a quotient and a remainder, where the remainder is less than n. This is a tact that most elementary school students learn when they are introduced to long division. In performing the division 1986 ÷ 97, you obtain a quotient of 20 and a remainder of 46;

i.e., $$1986 = (97)(20) + 46.$$

If two numbers, a and b, share the same remainder after dividing by n, we say that they are *congruent module n*, denoted a = b (mod n). For example, 13 = 38 (mod 4) because 13 = (5) (2) + 3 and 38 = (5) (7) + 3.

Modular Arithmetic: If n is a positive integer, we define the operations of addition n (× n) as follows. If, $a, b \in \mathbf{Z}$.

a + n b = remainder after a + b is divided by n, and

a × n b = remainder after a × b is divided by n.

Notes:

(a) The result of doing arithmetic module n is always an integer between 0 and n – 1 (Why?). This observation implies that {0, 1, ... n – 1} is closed under module n arithmetic.

(b) It is always true that a + nb = (a + b) (mod n) and a × nb = (a × b) (mod n). For example, $4 +_7 5 = 2 = 9$ (mod 7) and $4 \times_7 5 = 6 = 20$ (mod 7).

(c) We will use the notation $\mathbf{Z}_n$ to denote the set {0, 1, 2, ... n − 1}.

SUBSYSTEMS

The subsystem is a fundamental concept of algebra at the universal level.

Definition: *Subsystem. If [V; $*_1$,..., *n] is an algebraic system of a certain kind and W is a subset of V, then W is a subsystem of V if [W; $*_1$, ..., $*_n$] is an algebraic system of the same kind as V. The usual notation for "W is a subsystem of V" is W ≤ V.*

Since the definition of a subsystem is at the universal level, we can cite examples of the concept of subsystem at both the axiomatic and concrete level.

Example:

(a) (Axiomatic) If [G; *] is a group, and H is a subset of G, then H is a *subgroup* of G is [H; *] is a group.

(b) (Concrete) U = {−1, 1} is a subgroup of [**R***;]. Take the time now to write out the multiplication table of U and convince yourself that [U; .] is a group.

(c) (Concrete) The even integers, 2**Z** = {2k : k is an integer} is a subgroup of [**Z**; +]. Convince yourself of this fact.

(d) (Concrete) The set of nonnegative integers is not a subgroup of [**Z**; +]. All of the group axioms are true for this subset except one: no positive integer has a positive additive inverse. Therefore, the inverse property is not true. Note that every group axiom must be true for a subset to be a subgroup.

(e) (Axiomatic) It M is a monoid and P is a subset of M, then P is a *submonoid* of M if P is a monoid.

(f) (Concrete) If B* is the sent of strings of 0's and 1's of length zero or more with the operation of concatenation, then two submonoids of S are;

(i) The set of strings of even length, and

(ii) The set of strings that contain no 0's. The set of strings of length less than 100 is not a submonoid because it isn't closed under concatenation. Why isn't the set of strings of length 100 or more not a submonoid of B*?

For the remainder of this section, we will concentrate on the properties of subgroups. The first order of business is to establish a systematic way of determining whether a subset of a group is a subgroup.

ISOMORPHISMS

The following informal definition of isomorphic systems should be memorized. No matter show technical a discussion about isomorphic systems becomes, keep in mind that this is the essence of the concept.

Definition: *Isomorphic Systems/Isomorphism. Two algebraic systems are isomorphic if there exists a translation rule between them so that any true statement in one system can be translated to a true statement in the other system. The translation rule is called an isomorphism.*

Example 1:

Imagine that out are an eight-year-old child who has been reared in an English-speaking family, has moved to Greece, and has been placed in a Greek school.

Suppose that your new teacher asks the class to do an addition problem that has been written out in Greek, such as the one in the natural thing for you to do is to pull out your Greek-English/English-Greek dictionary and translate the Greek words to English. After you've solved the problem, you can consult the same dictionary to obtain the proper Greek word that the teacher wants. Although this is not the recommended method of learning a foreign language, it will surely yield the correct answer to the problem. Mathematically, we may say that the system of Greek integers with addition (kai) is isomorphic to English integers with addition (plus). The problem of translation between natural languages is more difficult than this though, because two complete natural languages are not isomorphic, or at least the isomorphism between them is not contained in a simple dictionary.

Example 2:

Pascal Sets. In this example, we will describe how set type variables, such as the ones in Pascal, can be implemented on a computer. We will describe the two systems first and then describe the isomorphism between them.

System 1: The Power Set of {1, 2, 3, 4, 5} with the Operation of Union ∪. For simplicity, we will only discuss union. However, the other operations are implemented in a similar way.

System 2: Strings of Five Bits of Computer Memory Together with An OR Gate. Individual bit values are either zero or one, so the elements of this system can be visualized as sequences of 5 zeros and ones. An OR gate, small piece of computer hardware that accepts two bits values at any one time and outputs either a zero or one, depending on the inputs. The output of an OR gate is one, except when the two bit values that it accepts are both zero, in which case the output is zero. The operation on this system actually consists of sequentialy inputting the values of two bit strings into the OR gate. The result will be a new string of five zeros and ones. An alternate method or operating in this system is to use five OR gates and to input corresponding pairs of bits from the input strings into the gates concurrently.

OPERATIONS

One of the first mathematical skills that we all learn is how to add a pair of positive integers. A young child soon recognizes that something is wrong if a sum has two values, particularly if his or her sum is different from the teacher's. In addition, it is unlikely that a child would consider assigning a non-positive value to the sum of two positive integers. In other words, at an early age we probably know that the sum of two positive integers is unique and belongs to the set of positive integers. This is what characterizes all operations on a set.

Definition: *Binary Operation. Let S be a non-empty set. A binary operation on s is a rule that assigns to each ordered pair of elements of S a unique element of S. In other words, a binary operation is a function from S × S into S.*

Example:

Union and intersection are both binary operations on the power set of any universe. Addition and multiplication are binary operators on the natural numbers. Addition and multiplication are binary operations on the set of 2 by 2 real matrices, $M_{2 \times 2}$ **(R)**. Division is a binary operation on some sets of numbers, like the positive reals. But on the integers ($1 \div 2 \notin$ **Z**) and even on the reals ($1 \div 0$ is not defined), division is not a binary operations.

Note:

(a) We stress that the image of each ordered pair must be in S. This requirement disqualifies subtraction on the natural numbers from consideration as a binary operation, since $1 - 2$ is not a natural number. Subtraction is a binary operation on the integers.

(b) *On Notation.* Despite the fact that a binary operation is a function, symbols, not letters, are used to name them. The most commonly used symbol for a binary operation is an asterisk, *. We will also use $ when a second symbol is needed.

(c) If * is a binary operation on S and a, b ∈ S, there are three common ways of denoting the image of the pair (a,b). They are:

*ab	a * b	Ab *
Prefix Form	Infix Form	Postfix Form

We are all familiar with infix form. For example, 2 + 3 is how everyone is taught to write the sum of 2 an 3. But notice how 2 = 3 was just described in the previous sentence! The word *sum* preceded 2 and 3. Orally, prefix form is quite natural to us. The prefix and postfix forms are superior to infix form in some respects. The algebraic expressions with more than one operation didn't need parentheses if they were in prefix or postfix form. Due to our familiarity with infix form, we will use it throughout the remainder of this book.

Some operations, such as negation of numbers and complementation of sets, are not binary, but binary operators.

Definition: Unary Operation. *Let S be a non-empty set. A unary operator on S is a rule that assigns to each element of S a unique element of S. In other words a unary operator is a function from S into S.*

TYPES OF BINARY OPERATIONS

1. **Commutative Operation :** *Binary operation o over a set G is said to be commutative, if for every pair of elements a, b ∈ G, aob = boa.*

   Thus addition and multiplication are commutative binary operations for natural numbers where as subtraction and division are not commutative because, for a, b ∈ N a – b = b – a and a ÷ = b ÷ a can not be true for every pair of natural numbers a and b.

   For example 5 – 3 ≠ 3 – 5 and 5 ÷ 3 ≠ 3 ÷ 5.

2. **Associative Operation :** *A binary operation o on a set G is called associative if (aob) oc = ao(boc) for all a, b, c ∈ G.*

   Evidently ordinary addition and multiplication are associative binary operations on the set of natural numbers, integers, rational numbers and real numbers. However, if we define

   $$aob = a - 3b \ \forall \ a, b \in R,$$

   then $(aob)oc \neq ao(boc)$ because

   $$(aob)oc = (aob) - 3c = (a - 3b) - 3c$$
   $$= a - 3b + 3c.$$

   and $ao\,(boc) = a - 3\,(boc) = a - 3\,(b - 3c)$

   $$= a - 3b + 9c$$

   Thus the operation defined as above is not associative.

3. **Distribution Operation** : Let * and o be two binary operations defined on a set G. Then the operation * is said to be left distributive with respect to operation o if a * (boc) = (a * b) o (a * c) for all a, b, c ∈ G and is said to be right distributive with respect to o if,

$$(boc) * a = (b * a) \text{ o } (c * a)$$

for all a, b, c ∈ G.

Whenever the operation * is left as well as right distributive, we simply say that * is distributive with respect to o.

Note: To prove that a binary operation in a set S obeys a law (commutative law) we must prove that elements of every ordered pair obeys the law, *i.e.*, the law must be proved by taking arbitrary elements. But to prove a binary operation is S does not obey a particular law, it is sufficient if we give a counter example. This method of proving the result is called the *proof by counter example.*

Example:

1. Let A is the set of even integers.
   (i) +, . are binary operations in A since for a, b ∈ A, a + b ∈ A and a b ∈ A.
   (ii) +, . are commutative in a since for a, b ∈ A, a + b = b + a and ab = ba.
   (iii) +, . are associative in A since for a, b, c ∈ A, (a + b) + c = a + (b + c) and a(bc) = (ab)c.
   (iv) . is distributive w.r.t. the operation + in A since for a, b, c ∈ A,

   a.(b + c) = a.b + a.c and

   (b + c).a = b.a + c.a
2. S is the set of all m × n matrices such that each element of any matrix is a complex number.

   Addition of matrices, denoted by +,- is a binary operation is S. Also (+) is commutative and associative is S.
3. S is the set of all vectors.
   (i) Addition of vectors, denoted by + is a binary operation is S. Also + is commutative and associative is S.
   (ii) Dot product of vectors denoted by ., is not a binary operation is S since for a, b ∈ S, a . b ∉ S.
   (iii) Cross product of vectors denoted by × is a binary operation in S since for a, b ∈ S, a × b ∈ S.

4. In N, the operation o defined by aob = $\frac{a+b}{ab}$ is not a binary operation.

COMPOSITION (OR OPERATION) TABLE

A binary operation in a finite set can completely be described by means of a table. This table is known as *composition table.* The composition table helps us to verify most of the properties satisfied by the binary operations.

This table can be formed as follows :

(i) Write the elements of the set (which are finite in number) in a row as well as in a column.

(ii) Write the element associated to the ordered pair (a_i, a_j) at the intersection of the row headed by a_i and the column headed by a_j. Thus

(ith entry on the left) . (jth entry on the top)

= (entry on the ith row and jth column intersect).

For example, the composition table for the group {(0, 1, 2, 3,4)} for the operation of addition is given below :

+	0	1	2	3	4
0	0	1	2	3	4
1	1	2	3	4	5
2	2	3	4	5	6
3	3	4	5	6	7
4	4	5	6	7	8

In the above example, the first element of the first row in the body of the table, 0 is obtained by adding the first element 0 of head row and the first element 0 of the head column. Similarly the third element of 4th row (5) is obtained by adding the third element 2 of the head row and the fourth element of the head column and so on.

An operation represented by the composition table will be binary, if every entry of the composition table belongs to the given set. It is to be noted that composition table contains all possible combinations of two elements of the set with respect to the operation.

Notes:

(i) If should be noted that the elements of the set should be written in the same order both in top border and left border of the table, while preparing the composition table.

(ii) Generally a table which defines a binary operation '.' on a set is called *multiplication table*, when the operation is '+' the table is called *an addition table.*

Composition Table for an Operation on Finite Sets (Cayley's Composition Table)

Sometimes an operation o on a finite set can conveniently be specified by a table called the composition table. The construction of the table is explained below :

Let $S = \{a_1, a_2, ..., a_i, a_j, ..., a_n\}$ be a finite set with n elements. Let a table with (n + 1) rows and (n + 1) columns be taken. Let the squares in the first row be filled in with o, $a_1, a_2, ..., a_n$ and the squares in the first column be filled in with o, $a_1, a_2, ..., a_n$. Let a_i $(1 \leq i \leq n)$ and $a_j (1 \leq j \leq n)$ be any two elements of S. Let the product a_i o a_j obtained by operating a_i with a_j be placed in the square which is at the intersection of the row headed by a_i and the column headed by a_j. Thus the following table be formed :

o	a_1	a_2	...	a_j	...	a_n
a_1	$a_1 o a_1$	$a_1 o a_2$	...	$a_1 o a_j$	...	$a_1 o a_n$
a_2	$a_2 o a_1$	$a_2 o a_2$	...	$a_2 o a_j$	...	$a_2 o a_n$
...	...	...	...	...	...	...
a_i	$a_i o a_1$	$a_i o a_2$	...	$a_i o a_j$	...	$a_i o a_n$
...	...	...	...	...	...	...
a_n	$a_n o a_1$	$a_n o a_2$	...	$a_n o a_j$	...	$a_n o a_n$

From the composition table we can infer the following laws :

(i) Closure Law : If all the products formed in the table are the elements of S, the 'o' is said to be a binary operation in S and S is said to be closed under the composition 'o'.

Otherwise, o is not a binary operation in S and the set S is not closed under the operation o.

(ii) Commutative Law : If the elements in every row are identical with the corresponding elements in the corresponding column, then the composition o is said to be commutative in S. Otherwise, the binary operation o is not commutative in S.

(iii) Associative Law : Also we can know from table whether the binary operation follows associative law or not.

Note: The diagonal through $a_1 o a_1$ and $a_n o a_n$ is called the leading diagonal in the table. If the elements in the table are symmetric about the leading diagonal then we infer that o is commutative in S.

Example:

Let S = {1, – 1, i, – i} and usual multiplication is the operation is S. Then we have the following composition table. We can clearly see that is a binary operation in S following commutative and associative laws.

•	1	–1	i	–i
1	1	–1	i	–i
–1	–1	1	–i	i
i	i	–i	–1	1
–i	–i	i	1	–1

ALGEBRAIC STRUCTURE

A non-empty set G equipped with one or more binary operations in called on *algebraic structure or an algebraic system.*

If o is a binary operation on G; then the algebraic structure is written as (G, o), e.g., (N, +), (Q, –), (R, +) are algebraic structures.

PROPERTIES OF MODULAR ARITHMETIC ON ZN

Addition module n is always commutative an associative; 0 is the identity for + n and every element of $\mathbf{Z}_n$ has an additive inverse.

Multiplication module n is always commutative and associative, and 1 is the identity for × .

Multiplication module n is distributive (both right and left) over addition module n.

Theorem:

If $a \in Zn$, $a \neq 0$, then $-a = n - a$.

Proof:

a + n(n – a) = (a + (n – a) (mod n) = n (mod n). The only, element of Z_n that is congruent to n is 0. #

Note: The algebraic properties of + n and × n on Z_n are identical to the properties of addition and multiplication on **Z**.

The group Z_n: For each n ≥ 1, $[Z_n; +_n]$ is a group. Henceforth, we will use the notation Zn when referring to this group contains the tables for Z_1 through Z_5.

Example:

(a) We are all somewhat familiar with $\mathbf{Z}_{12}$ since the hours of the day are counted using this group, except for the fact that 12 is used in place of 0. If someone started a four-hour trip at 10 o'clock, she would arrive at $10 -_{12} 4 = 2$ o'clock. If a satellite orbits the earth every two hours and starts its first orbit at 5 o' clock, it would end its first orbit at 5 + 122 = 7 o'clock. Its seventh orbit would end at at 5 + 12 7 (2) = 7 o'clock.

$+_1$	0
0	0

$+_2$	0	1
0	0	1
1	1	0

$+_3$	0	1	2
0	0	1	2
1	1	2	0
2	2	0	1

$+_4$	0	1	2	3
0	0	1	2	3
1	1	2	3	0
2	2	3	0	1
3	3	0	1	0

$+_5$	0	1	2	3	4
0	0	1	2	3	4
1	1	2	3	4	0
2	2	3	4	5	0
3	3	4	0	1	2
4	4	0	1	2	3

(b) Virtually all computers represent unsigned integers in binary form with a fixed number of digits. A very small computer might reserve seven bits to store the value of an integer. There are only 2^7 different values that can be stored in seven bits. Since the smallest value is 0, represented as 0000000, the maximum value will be $2^7 - 1 = 127$, represented as 11111111. When a command is given to add two integer values, and the two values have a sum of 128 or more, overflow occurs. For example, if we try to add 56 and 95, the sum is an eight-digit binary integer 10010111. One common procedure is to retain the seven lowest-ordered digits. The result of adding 56 and 95 could be $0010111_{two} = 23 = (56 + 95)$ (*mod* 128). Integer arithmetic with this computer would actually be module 128 arithmetic.

GROUPOID

A groupoid is a pair (G, o) consisting of a non-empty set G, called the carrier and a binary operation o in G.

Examples:

(i) Let G = {1, 2} and let o be the binary operation in G defined as follows :

1o1 = 1, 1o2 = 2, 2o1 = 1, 2o2 = 2.

Then (G, o) is a groupoid.

(ii) Let o be the binary operation in Q, the rational numbers defined by aob = a + b + ab. Then (Q, o) is a groupoid, because for every pair of rational numbers a and b, aob defines a unique rational number a + b + ab.

Equality of Groupoids : Two groupoids are equal if and only if they have the same carriers and the same binary operation.

Order of a Groupoid : The *order* of a groupoid (G, o) is the number of elements in G and is denoted by |G|, (G, o) is *infinite* if |G| is infinite and *finite* if |G| is finite.

GROUPS

We begin by considering two familiar algebraic systems, each of which is a group, with a view to pointing out those features common to both which are set forth abstractly in the general concept of a group.

We first observe that the set R of all real numbers, together with the operation of ordinary addition, has the following properties: the sum of any two numbers in R is a number in R (R is closed under addition); if x, y, z are any three numbers in R, then x + (y + z) = (x + y) + 2 (addition is associative); there is present in R a special number, namely 0, with the property that x + 0 = 0 + x = x for every x in R (R contains an additive identity element); and to each number x in R there corresponds another number in R, its negative –x, with the property that x + (–z) = (–x) + x = 0 (R contains additive inverses).

It is equally clear that the set P of all positive real numbers, together with the operation of ordinary multiplication, has the following corresponding properties: the product of any two numbers in P is a number in P (P is closed under multiplication); if x, y, z are any three numbers in P, then x{yz) = (xy)z (multiplication is associative); there is present in P a special number, namely 1, with the property that x1 = 1x = x for every a; in P (P contains a multiplicative identity element); and to each number x in P there corresponds another number in P, its reciprocal $1/x = x^{-1}$, with the property that $xx^{-1} = x^{-1}x = 1$ (P contains multiplicative inverses).

Each of these systems plainly possesses many properties other than those we have mentioned. We ignore all such properties and concentrate our attention solely on the ones we have listed. Let us now consciously disregard the concrete nature of the elements composing the above sets and the familiar character of the algebraic operations involved. What remains in each case is a non-empty set which is closed under an operation possessing certain formal properties, and apart from notation and terminology, these properties are identical in the two systems. The concept of a group is a distillation of the common structural form of these and many other similar systems. The definition is as follows.

Definition : *An algebraic structure (G, o) where G is a non-empty set with a binary operation 'o' defined on it is said to be a group, if the binary operation satisfies the following axiom (called group axioms).*

(G_1) Closure axiom : G is closed under the operation o, *i.e.*, $a \circ b \in G$, for all

$a, b \in G$.

(G_2) Associative axiom : The binary operation o is associative *i.e.*,

$(a \circ b) \circ c = a \circ (b \circ c) \ \forall \ a, b, c \in G$.

(G_3) Identity axiom : There exists an element $e \in G$ such that

$e \circ a = a \circ e = a \ \forall \ a \in G$.

The element e is called the identity of 'o' in G.

(G_4) Inverse axiom : Each element of G possesses inverse, *i.e.* for each element $a \in G$, there exists an element $b \in G$ such that

The element b is then called the inverse of a with respect to 'o' and we write $b = a^{-1}$. Thus a^{-1} is an element of G such that

$$a^{-1} \circ a = a \circ a^{-1} = e.$$

Abelian Group or Commutative Group Definition

A group (G, o) is said to be abelian or commutative if the composition 'o' is commutative, i.e., if,

$$a \circ b = b \circ a \ \forall \ a, b \in G.$$

A group which is not abelian is called non-abelian.

Examples:

(i) The structures (N, +) and (N, ×) are not groups *i.e.*, the set of natural numbers considered with the addition composition or me multiplication composition, does not form a group. For, the postulate (G_3) and (G_4) in the former case, and (G_4) in the latter case, are not satisfied.

(ii) The structure (Z, +) is a group, *i.e.*, the set of integers with the addition composition is a group. This is so because addition in numbers is associative, the additive identity O belongs to Z, and the inverse of every element a, viz. –a belongs to Z. This is known as additive group of integers.

The structure (Z, ×), *i.e.* the set of integers with the multiplication composition does not form a group, as the axiom (G_4) is not satisfied.

(iii) The structures (Q, +), (R, +), (C, +) are all groups, *i.e.*, the sets of rational numbers, real numbers, complex numbers, each with the additive composition, form a group.

But the same sets with the multiplication composition do not form a group, for the multiplicative inverse of the number zero does not exist in any of them.

(iv) The structure (Q_0, ×) is group, where Q_0 is the set of non-zero rational numbers. This is so because the operation is associative, the multiplicative identity 1 belongs to Q_0 and the multiplicative inverse of every element a in the set is 1/a, which also belongs to Q_0. This is known as the *multiplicative group* of non-zero rationals.

Obviously (R_0 ×) and (C_0, ×) are groups, where R_0 and Co) are respectively the sets of non-zero real numbers and non-zero complex numbers.

(v) The structure (Q^+, ×) is a group, where Q^+ is the set of positive rational numbers: It can easily be seen that all the postulates of a group are satisfied.

Similarly, the structure (R^+, ×) is a group, where R^+ is the set of positive real numbers.

(vi) The, groups in (ii), (iii), (iv) and (v) above are all *abelian groups*, since addition and multiplication are both commutative operations is numbers.

Finite and Infinite Groups

If a group contains a finite number of distinct elements, it is called *finite group* otherwise an infinite group.

In other words, a group (G, o) is said to be finite or infinite according as the underlying set G is finite or infinite.

Order of a Group

The number of elements in a finite group is called *order* of the group. An infinite group is said to be of *infinite order.*

Note : It should be noted that the smallest group for a given composition is the set {e} consisting of the identity element e alone.

GROUP TABLES

The composition tables are useful in examining the following axioms in the manner explained below:

1. **Closure Property :** If all the elements of the table belong to the set G (say) then G is closed under the Composition o (say). If any of the elements of the table does not belong to the set, the set is not closed.
2. **Existence of Identity :** The element (in the vertical column) to the left of the row identical to the top row (border row) is called an identity element in the G with respect to operation 'o'.
3. **Existence of Inverse :** If we mark the identity elements in the table then the element at the top of the column passing through the identity element is the inverse of the element in the extreme left of the row passing through the identity element and vice-versa.
4. **Commutativity :** It the table is such that the entries in every row coincide with the corresponding entries in the corresponding column *i.e.*, the composition table is symmetrical about the principal or main diagonal, the composition is said to have satisfied the commutative axiom otherwise it is not commutative.

The process will be more clear with the help of following illustrative examples.

GENERAL PROPERTIES OF GROUPS

Theorem 1:

G is a group with binary operation o and if a and b are any elements of G, then the linear equations

$$a \circ x = b \text{ and } y \circ a = b$$

have unique solutions in G.

Proof:

Now $a \in G \Rightarrow a^{-1} \in G,$

and $a^{-1} \in G, b \in G \Rightarrow a^{-1} \circ b \in G.$

Substituting $a^{-1} \circ b$ for x in the equation $a \circ x = b$, we obtain

$$a \circ (a^{-1} \circ b) = b$$

$$\Rightarrow (a \circ a^{-1}) \circ b = b$$

$$\Rightarrow e \circ b = b$$

$$b = b \qquad [\because \text{ e is the identity}]$$

Thus $x = a^{-1} ob$ is a solution of the equation $aox = b$.

To show that the solution is unique let us suppose that the equation $aox = b$ has two solutions given by

$$x = x_1 \text{ and } x = x_2$$

Then $aox_1 = b$ and $aox_2 = b$

$\Rightarrow \quad aox_1 = aox_2 = b$

$\Rightarrow \quad x_1 = x_2$ (by left cancellation law)

In a similar manner, we can prove that the equation

$$yoa = b$$

has the unique solution $y = b \text{ o } a^{-1}$.

Theorem 2(a):

The inverse of the product of two elements of a group G is the product of the inverse taken in the reverse order i.e.,

$$(a \text{ o } b)^{-1} = b^{-1} \text{ o } a^{-1} \;\forall\; a, b \in G$$

Proof:

Let us suppose a and b are any two elements of G. If a^{-1} and b^{-1} are inverses of a and b respectively, then

$a^{-1} \text{ o } a = e = a \text{ o } a^{-1}$ (e being the identity element)

and $b^{-1} \text{ o } b = e = b \text{ o } b^{-1}$

Now, $(a \text{ o } b) \text{ o } (b^{-1} \text{ o } a^{-1}) = [(a \text{ o } b) \text{ o } b^{-1}]oa^{-1}$ (by associativity)

$= [a \text{ o } (b \text{ o } b^{-1})] \text{ o } a^{-1}$ (by associativity)

$= [a \text{ o } (b \text{ o } b^{-1})] \text{ o } a^{-1}$ (by associativity)

$= (a \text{ o } e) \text{ o } a^{-1}$ $[\because b \text{ o } b^{-1} = e]$

$= a \text{ o } a^{-1}$ $[\because a \text{ o } e = a]$

$= e.$ $[\because a \text{ o } a^{-1} = e]$

Also $(b^{-1} \text{ o } a^{-1}) \text{ o } (a \text{ o } b) = b^{-1} \text{ o } [a^{-1} \text{ o } (a \text{ o } b)]$ (by associativity)

$= b^{-1} \text{ o } [(a^{-1} \text{ o } a) \text{ o } b]$

$= b^{-1} \text{ o } (e \text{ o } b)$ $[\because a^{-1} \text{ o } a = e]$

$= b^{-1} \text{ o } b$ $[\because e \text{ o } b = b]$

$= e.$

Hence, we have

$$(b^{-1} \text{ o } a^{-1}) \text{ o } (a \text{ o } b) = e = (a \text{ o } b) \text{ o } (b^{-1} \text{ o } a^{-1})$$

Therefore, by definition of inverse, we have

$$(a \circ b)^{-1} = b^{-1} \circ a^{-1}$$

This theorem can be generalised as :

if a, b, c, ... k, 1, m $\in$ G, then

$$(a \circ b \circ c \circ \ldots k \circ l \circ m)^{-1} = m^{-1} \circ l^{-1} \circ k^{-1} \circ \ldots c^{-1} \circ b^{-1} \circ a^{-1}$$

Theorem 2(b):

Cancellation laws hold good in a group, i.e., if a, b, c, are any elements of G, then

$a \circ b = a \circ c \Rightarrow b = c$ *(left cancellation law)*

and $b \circ a = c \circ a \Rightarrow b = c$ *(right cancellation law)*

Proof:

Let a $\in$ G. Then

$a \in G \Rightarrow G$ such that $a^{-1} \circ a = e = a \circ a^{-1}$,

where e is the identity element

Now, let us assume that

$a \circ b = a \circ c$

then $a \circ b = a \circ c \Rightarrow a^{-1} \circ (a \circ b) = a^{-1} (a \circ c)$

$\Rightarrow$ $(a^{-1} \circ a) \circ b = (a^{-1} \circ a) \circ c$ (by associative law)

$\Rightarrow$ $e \circ b = e \circ c$ ($\because a^{-1} \circ a = e$)

$\Rightarrow$ $b = c$.

Similarly, $b \circ a = c \circ a$

$\Rightarrow$ $(b \circ a) \circ a^{-1} = (c \circ a) \circ a^{-1}$

$\Rightarrow$ $b \circ (a \circ a^{-1}) = c \circ (a \circ a^{-1})$

$\Rightarrow$ $b \circ e = c \circ e$

$\Rightarrow$ $b = c$.

Theorem 3:

The identity element of a group is unique.

Proof:

Let us suppose e and e' are two identity elements of group G, with respect to operation o.

Then $e \circ e' = e$ if e' is identity.

and $e \circ e' = e'$ if e is identity.

But $e \circ e'$ is unique element of G, therefore,

$$e \text{ o } e' = e \text{ and } e \text{ o } e' = e' \Rightarrow e = e'$$

Hence the identity element in a group is unique.

Theorem 4(a):

The inverse of each element of a group is unique, i.e., in a group G with operation o for every $a \in G$, there is only one element a^{-1} such that

$$a^{-1} \text{ o } a = a \text{ o } a^{-1} = e, \text{ e being the identity.}$$

Proof:

Let a be any element of a group **G** and let e be the identity element. Suppose there exist a^{-1} and a' two inverses of a in **G** then

$$a^{-1} \text{ o } a = e = a \text{ o } a^{-1}$$

and $$a' \text{ o } a = e = a' \text{ o } a$$

Now, we have

$$a^{-1} \text{ o } (a \text{ o } a') = a^{-1} \text{ o } e \qquad \text{(since } a \text{ o } a' = e)$$

$$= a^{-1} \qquad (\because e \text{ is identity})$$

Also, $$(a^{-1} \text{ o } a) \text{ o } a' = e \text{ o } a' \qquad (\because a^{-1} \text{ o } a = e)$$

$$= a' \qquad (\because e \text{ is identity})$$

But $a^{-1} \text{ o } (a \text{ o } a') = (a^{-1} \text{ o } a) \text{ o } a'$ as in a group composition is associative

$\therefore \ a^{-1} = a^{1}$

Theorem 4(b):

If the inverse of a is a^{-1} then the inverse a^{-1} is a, i.e., $(a^{-1})^{-1} = a$.

Proof:

If e is the identity element, we have $a^{-1} \text{ o } a = e$ (by definition of inverse)

$\Rightarrow \ (a^{-1})^{-1} \text{ o } (a^{-1} \text{ o } a) = (a^{-1})^{-1} \text{ oe} \qquad [\because a^{-1} \in G \Rightarrow (a^{-1})^{-1} \in G]$

$\Rightarrow \ [(a^{-1})^{-1} \text{ o } a^{-1}] \text{ o } a = (a^{-1})^{-1}$

[$\because$ Composition in G is associative and e is identity element]

$\Rightarrow \ e \text{ o } a = (a^{-1})^{-1} \Rightarrow a = (a^{-1})^{-1} \Rightarrow \ (a^{-1})^{-1} = a.$

Theorem 4(c):

If corresponding to any element $a \in G$; there is an element which satisfies one of the conditions

$$a + 0_a = a \text{ or } 0_a + a = a$$

then it is necessary that $O_a = 0$, where 0 is the identity element of the group.

Proof:

Since 0 is the identity element,

We have $a + 0 = a$ ...(i)

Also, it is given that

$a + 0_a = a$ ...(ii)

Hence, from (i) and (ii)

$a + 0_a = a + 0$

or $0_a = 0$ (by left cancellation law)

Again, we have $0 + a = a$ ...(iii)

and $0_a + a = a$ (given ...(iv)

Hence, from (iii) and (iv), we get

$0_a + a = 0 + a$

so that $0_a = 0$ (by right cancellation law.)

SOME DEFINITIONS

(i) If for every $a \in G$, there exists an element e in G, such that **eoa** = a, then e is called, the left identity.

(ii) If for every $a \in$ **G**, there exists an element e in **G** such that **aoe** = a, then e is called the right identity.

(iii) If for an $a \in G$ there exists an element a^{-1} in G such that $a \circ a^{-1} = e$, then a^{-1} is called the left inverse of a.

(iv) If for an $a \in G$ there exists an element a^{-1} in G such that $a \circ a^{-1} = e$, then a^{-1} is called the right inverse of a.

Theorem 1:

The left inverse of an element is also its right inverse, i.e.,

$a^{-1} \circ a = e = a \circ a^{-1}$

Proof:

Now, $a^{-1} \circ (a \circ a^{-1}) = (a^{-1} \circ a) \circ a^{-1}$ (by associative law)

$= a^{-1} \circ e$ [by theorem 1]

Thus $a \circ (a \circ a^{-1}) = a^{-1} \circ e$

$\therefore$ $a \circ a^{-1} = e$ [by left cancellation law]

Hence a^{-1} is also the right inverse of a.

Theorem 2:

The left identity is also the right identity ie.,

$$e \circ a = a = a \circ e \ \forall \ a \in G$$

Proof:

If a^{-1} be the left inverse of a,

then $a^{-1} \circ (a \circ e) = (a^{-1} \circ a) \circ e$ (By associative law)

or $a^{-1} \circ (a \circ e) = e \circ e$ (by definition of left inverse)

$= e$

$= a^{-1} \circ a$

Thus $a^{-1} \circ (a \circ e) = a^{-1} \circ a$

$\therefore$ $a \circ e = a$ [by left cancellation law]

Hence e is also the right identity element.

An Alternative Definition for a Group

A set G with a binary composition denoted multiplicatively is a group if

(i) the composition is associative.

(ii) for every pair of elements a, b $\in$ G, the equations a x = b and ya = b have unique solutions in G.

Proof:

Binary operation implies that the set **G**, under consideration is closed under the operation. Now to prove set **G** a group we have to show that the left identity exists, and each element of **G** possesses left inverse with respect to the operation under consideration.

It is given that for every pair of elements a, b $\in$ **G** the equation ya = b has a solution in **G**. Therefore, if a $\in$ **G**, then taking b = a, we observe that there exists an element, say **e** $\in$ **G** such that

$$ea = a \quad ...(i)$$

Now, let us suppose that b is any arbitrary element of **G**.

Therefore, there exists x $\in$ G such that

$$ax = b \quad ...(ii)$$

Thus $b = ax \Rightarrow eb = e\,(ax)$

$\Rightarrow$ $eb = (ea)\,x$ (associative law)

$\Rightarrow$ $eb = ax$ [from (i)]

$\Rightarrow$ $eb = b$ [from (ii)]

Therefore $\exists$ e $\in$ G such that

$$eb = b \ \forall \ b \in G$$

$\therefore$ e is the left identity,

Now, let b = e $\in$ G be an element.

$\therefore ya = b \Rightarrow ya = e$

$\Rightarrow$ y is inverse of a in G.

Let, $y = a^{-1}$ such that

$a^{-1} a = e$ then $a^{-1} \in G$ as ya = e has got solution in G.

Thus, a^{-1} is the left inverse of a in G. Therefore each element of G possesses left inverse.

Hence, G is a group for the given composition if the postulates (i) and (ii) are satisfied.

SEMI-GROUP

An algebraic structure (S, o) is called a semi group if the binary operation o is associative in S.

Examples:

(i) (N, +) is a semigroup. For a, b $\in$ N

$\Rightarrow$ a + b $\in$ N and a, b, c $\in$ N,

$\Rightarrow$ $(a + b) + c = a + (b + c)$

(ii) (R, +) is a semi-group. For a, b $\in$ R

$\Rightarrow$ a + b $\in$ R and a, b, c $\in$ R

$\Rightarrow$ $(a + b) + c = a + (b + c)$

(iii) (P(S), $\cap$) is a semi-group where P(S) is the power set of non-empty set S.

(iv) Q is the set of rational number, o is a binary operation defined on Q such that aob = a – b + ab for a, b $\in$ Q. (Q, o) is not a semi-group.

For a, b, c $\in$ Q,

$$\begin{aligned}(aob)oc &= (aob) - c + (aob)c\\ &= a - b + ab - c + (a - b + ab)c\\ &= a - b + ab - c + ac - bc + abc\end{aligned}$$

$$\begin{aligned}\text{and } ao(boc) &= a - (boc) + a\,(boc)\\ &= a - (b - c + bc) + a\,(b - c + bc)\\ &= a - b + c - bc + ab - ac + abc\end{aligned}$$

and (aob)oc $\neq$ ao(boc).

PRODUCT AND QUOTIENTS OF SEMI-GROUPS

Product : If (S, o) and (T, *) are semi-groups, then (S × T, *') is a semi-group, where *' is defined by

$$(s_1, t_1) *' (s_2, t_2) = (s_1 \text{ o } s_2, t_1 * t_2).$$

Quotient : Equivalence relation R on the semi-group (S, o) determines) a partition of S. We let [a] = R(a) be the equivalence class containing a and S/R denote the set of all equivalence classes. Then S/R is a semi-group and is called the *quotient semi-group* or *factor semi-group.*

Congruence Relation : An equivalence relation R on the semi-group (S, o) is called a congruence relation if

a R a' and b R b' imply (aob) R (a' o b').

Theorem 1:

*Let R be a congruence relation on the semi-group (S, o). Consider the relation * from S/k × S/R to S/R in which the ordered pair ([a], [b]) is, for a and b in S, related to [a o b].*

(a) * is a function from S/R × S/R to S/R, i.e., [a] * [b] = [a o b]

(b) (S/R, *) is a semi-group.

Proof:

(a) Suppose that ([a], [b]) = ([a'], [b']). Then a R a' and b R b', some must have a o b R a' o b', since R is a congruence relation. Thus [a o b] = [a' o b']; that is, * is a function. This means that * is a binary operation on S/R.

(b) Now,

[a] * ([b] * [c]) = [a] * [b o c]

= [a o (b o c)]

= [(a o b) o c] as (S, o) is a semi-group.

= [a o b] * [c]

= ([a] * [b]) * [c].

Hence S/R is a semi-group.

Theorem 2:

*Let R be a congruence relation on a semi-group (S, o), and let (S/R, *) be the corresponding quotient semi-group. Then the function $f_R : S \to S/R$ defined by*

$$f_R(a) = [a]$$

is an onto homomorphism, called the natural homomorphism.

Proof:

if [a] $\in$ S/k, then $f_R(a)$ = [a], so f_R is an onto function. Moreover, if a and b are elements of S, then

$$f_R\ (a \circ b) = [a \circ b] = [a] * [b]$$
$$= f_R(a) * f_R(b),$$

so f_R is a homomorphism.

DEFINITION OF A GROUP BASED UPON LEFT AXIOMS

*Let G be a non-empty set equipped with a binary operation denoted by *, then this algebraic structure (G, *) is a group if the binary operation * satisfies the following postulates :*

1. **Closure property,** *i.e.*, ab $\in$ G $\forall$ a, b $\in$ G
2. **Associativity,** *i.e.*, (ab) c = a (bc) $\forall$ a, b, c $\in$ G
3. **Existence of Left Identity :** *There exists an element e $\in$ G such that ea = a $\forall$ a $\in$ G. The element is called the left identity.*
4. **Existence of Left Inverse.** *Each element of G possesses left inverse. In other words a $\in$ G $\Rightarrow$ there exists an element $a^{-1} \in G$ such that $a^{-1}\,a = e$. The* element a^{-1} is the left inverse of a.

Proof:

This definition of a group and the classical definition of a group are equivalent, obviously if the postulates of a group is hold good, the postulates of a group given in this definition will also hold good.

If the postulate of a group given in this definition hold good, then the postulates will also hold good if starting with left axioms we prove that the left identity is also the right identity and the left inverse of an element is also the right inverse.

MODULO SYSTEM

It is of common experience that railway time-table is fixed with the provision of 24 hours in a day and night. When we say that a particular train is arriving at 15 hours, it implies that the train will arrive at 3 p.m. according to our watch. Thus all the timing starting from 12 to 23 hours correspond to one of 0, 1, 3 ... 11 O'clock as indicated in watches. In other words all integers from 12 to 23 are equivalent to one or the other of integers 0, 1, 2, 3, ...11 with modulo 12. In saying like this the integers in question are divided into 12 classes.

In the manner described above the integer could be divided into 2 classes, or 5 classes or m (m being a positive integer) classes and then we would have written mod 2 or mod 5 or mod m. This system of representing integers is called *modulo system.*

Addition Modulo m

We shall now define a new type of addition known as "addition modulo m" and written as $a +_m b$ where a and b are any integers and m is a fixed positive integer.

By definition, we have

$$a +_m b = r, 0 \leq r < m$$

where r is the least non-negative remainder when a + b, *i.e.*, the ordinary sum of a and b, is divided by m.

For example $5 +_6 3 = 2$, since $5 + 3 = 8 = 1\ (6) + 2$, *i.e.*, 2 is the least non-negative remainder when 5 + 3 is divided by 6. Similarly, $5 +_7 2 = 0$,

$$4 +_3 2 = 0;\ 3 +_3 1 = 1,\ 15 +_5 7 = 2.$$

Thus to find $a +_m b$, we add a and b in the ordinary way and then from the sum, we remove integral multiples of m in such a way that the remainder r is either 0 or a positive integer lessman m.

When a and b are two integers such that a – b is divisible by a fixed positive integer m, then we write

$$a = b \pmod{m}$$

which is read as "a is congruent to b modulo m."

Thus a = b (mod m) if a – b is divisible by m. For example 13 = 3 (mod 5) since 13 – 3 = 10 is divisible by 5, 5 = 5 (mod 5), 16 = 4 (mod 6); –20 = 4 (mod 6).

Multiplication Modulo p

We shall now define a new type of multiplication known as "multiplication modulo p" and written as $a \times_p b$ where a and b are any integers and p is a fixed positive integer.

By definition we have

$$a \times_p b = r, 0 \leq r \leq p,$$

where r is the least non-negative remainder when ab, *i.e.*, the ordinary product of a and b, is divided by p.

For example $4 \times_7 2 = 1$,

since $4 \times 2 = 8 = 1\ (7) + 1$.

It can be easily shown that if a = b (mod p), men $a \times_p C = b \times_p C$

Additive Group of Integers Modulo m

The set G = (0, 1, 2, ... m –1} of first m non-negative integers is a group. The composition being addition reduced modulo m.

Closure Property : We have by definition of addition modulo m,

$$a + mb = r$$

where r is the lest non-negative remainder when the ordinary sum a + b is divided by m. Obviously $o \le r \le m - 1$. Therefore for all a b $\in$ G we have a + mb $\in$ G and thus G is closed with respect to the composition addition modulo m.

Associative Property : Let a, b, c be any arbitrary elements in G. Then

$(a + b) +_m c = (a +_m b) +_m c$ $\qquad [\because b +_m c = b + c \pmod m]$

= least non-negative remainder when a + (b + c) is divisible by m

= least non-negative remainder when (a + b) + c is divided by m, since

$$a + (b + c) = (a + b) + c$$

$= (a + b) +_m c$ [by definition of ^+m]

$= (a +_m b) +_m c$ $\qquad [\because a + b = a +_m b \pmod m]$

$\therefore$ '$+_m$' is an associative composition.

Existence of Identity Element : We have $0 \in$ G. Also, if a is any element of G, then $0 +_m a = a = a +_m 0$. Therefore 0 is the identity element.

Existence of Inverse : The inverse of 0 is 0 itself. If $r \in$ G and $r \neq 0$, then

$m - r \in$ G. Also $(m - r) +_m r = 0 = r +_m (m - r)$. Therefore (m – r) is the inverse of r.

Commutative Property : The composition '+m' is commutative also.

Since

$a +_m b$ = least-non-negative remainder when a + b is divided by m

= least non-negative remainder when b + a is divided by m

$= b +_m a.$

The set **G** contains" m elements.

Hence $(G, +_m)$ is a finite abelian group of order m.

Multiplicative Group of Integers Modulo p where p is Prime

*The set **G** of (p – 1) integers 1, 2, 3, ...p – 1, p being prime, is finite abelian group of order p – 1 the composition being multiplication modulo p.*

Let $G = \{1, 2, 3, \ldots p-1\}$ where p is prime.

Closure Property : Let a and b be any elements of G. Then

$1 \le a \le p-1$, $1 \le b \le p-1$. Now by definition. $a \times_p b = r$ where r is the least non-negative remainder when the ordinary product a b is divided by p. Since p is prime, therefore a b is not exactly divisible by p. Therefore r can not be zero and we shall have $1 \le r \le p-1$. Thus $a \times_p b \in G \ \forall\ a, b \in G$. Hence the closure axiom is satisfied.

Associative Law : Let a, b, c, be any arbitrary elements of G.

Then $a \times_p (b \times_p c) = a \times_p (bc)$ $\quad [\because b \times_p c = bc \pmod p]$

= least non-negative remainder when a(bc) is divided by p

= least non-negative remainder when ab(c) is divided by p

$= (ab) \times_p c$

$= (a \times_p b) \times_p c$ $\quad [\because ab = a \times_p b \pmod p]$

$\therefore$ 'x_p' is an associative composition.

Existence of left identity : We have $1 \in G$. Also if a is any element of G, then $1 \times_p a = a$. Therefore 1 is the left identity.

Existence of left inverse : Lets be any member of G.

Then $1 < s < p - 1$.

Let us consider the following (p – 1) products :

$$1 \times_p s,\ 2 \times_p s,\ 3 \times_p s,\ \ldots\ (p-n) \times_p s.$$

All these are elements of G. Also no two of these can be equal as shown below :

Let i and j be two unequal integer such that

$$1 \le i \le p-1,\ 1 \le j \le p-1 \text{ and } i > j.$$

Then $1 \times_p s = j \times_p s$

$\Rightarrow$ is and js leave the same least non-negative remainder when divided by p

$\Rightarrow$ is -j s is divisible by p.

$\Rightarrow$ (i - j) s is divisible by p.

Since $1 \le (i - j) < p - 1$; $1 \le s \le p - 1$ and p is prime, therefore (i – j) s can not be divided by p.

$\therefore\ i \times_p s \ne j \times_p s.$

Thus $1 \times_p s$, $2 \times_p s$, ... $(p-1) \times_p s$ are (p – 1) distinct elements of the set G. Therefore one of these elements must be equal to 1.

Let Then $s' \times_p s = 1$. Then s' is the left inverse of s.

Commutative Law : The composition '$\times_p$' is commutative, since

$a \times_p b$ = least non-negative remainder when ab is divisible by p

= least non-negative remainder when ba is divided by p

$= b \times_p a$

$\therefore$ $(G, \times_p)$ is a finite abelian group of order $p - 1$.

Theorem 1:

The residue classes modulo form a finite group with respect to addition of residue classes.

Proof:

Let **G** be the set of residue classes (mod m), then

$$\{G = \{\{0\}, \{1\}, ...\{r_1\}, ...\{r_2\}, ... \{m-1\}\}$$

or $$G = \{0, 1, 2, ... r_1, ... r_2 .... m - 1 \ (\text{mod } m)\}$$

Closure axiom : $\{r_1\} + \{r_2\} = \{r_1 + r_2\} = \{r\} \in G$,

where r is the least positive integer obtained as remainder when $r_1 + r_2$ is divided by m $(0 \leq r < m)$.

Thus, the closure axiom is satisfied.

Associative axiom : The addition is associative.

Identity axiom : $\{0\} \in (G)$ and $\{0\} + \{r\} = \{r\}$. Hence the identity for addition is $\{0\}$.

Inverse axiom : Since $\{m - r\} + \{r\} = \{m\} = \{0\}$, the additive inverse of the element (r) is $(m - r)$.

Hence **G** is a finite group with respect to addition modulo m.

Theorem 2:

The set of non-zero residue classes modulo p. Where p is a prime, forms a group with respect to multiplication of residue classes.

Proof:

Let $\mathbf{I} = \{...., -3, -2, -1, 0, 1, 2, 3, ...\}$ be the set of integers. Let $a \in \mathbf{I}$, then $\{a\}$ is residue class modulo p of **I**, if $\{a\} = \{x : x \in \mathbf{I}$ and $x - a$ is divisible by p$\}$.

If p|a then $\{a\} = \{0\}$ which is called the zero residue class. Let **G** be the set of non-zero residue classes mod p(p being prime) then

$$G = \{1, 2, 3, ... (p - 1)\}$$

Closure axiom : Let $r_1, r_2 \in$ **G** then

$$r_1 \cdot r_2 = r \pmod p$$

where, r is the least non-negative integer such that $0 \leq r \leq p - 1$ obtained after dividing r_1, r_2 by p.

Also, since p is prime r_1, r_2 is not divisible by p. Hence r cannot be zero.

Hence, $r_1 \cdot r_2 = r = G$.

Thus closure axiom is satisfied

Associative axiom : Multiplication of residue classes is associative.

Existence of Identity : $I \in G$ and $a \cdot 1 = 1\ a = a\ \forall\ a \in G$.

Therefore, 1 is the identity element in **G** with respect to multiplication.

Existence of Inverse : Let $s \in G$ then $1 \leq s \leq p - 1$. Let us consider following (p – 1) elements.

$$1.s,\ 2 \cdot s,\ 3 \cdot s \ldots,\ (p - 1) \cdot s$$

All these elements are elements of **G** because the closure law is true. All these elements are distinct as otherwise if

$$i.s = j.s \text{ for } i \neq j \text{ and } j, j \in G$$

then $i.s = j.s \Rightarrow i.s - j.s$ is divisible by p

$\Rightarrow$ (i – j).s is divisibly by p

$\Rightarrow$ (i – j) is divisibly by p $[\because 1 \leq s < p - 1]$

$\Rightarrow$ i = j which is contrary to our assumption that $i \neq j$,

Therefore above (p – 1) elements are the same as the elements of **G**. Hence some one of them should be 1 also. Let $s'.s = 1$ where $1 \leq s' \leq p - 1$. Hence s' is inverse of s. Hence inverse axiom is also satisfied.

$\therefore$ G is a group under multiplication mod p.

Note: Since $r.s = s.r\ \forall\ r, s \in$ **G**.

G is finite abelian group of order (p — 1).

MONOIDS

Recall the definition of a monoid:

Definition: *Monoid. A monoid is a set M together with a binary operation * with the properties.*

(a) * is associative: $(a * b) * c = a * (b * c)$ for all $a, b, c \in M$ and

(b) * has an identity: there exists $e \in M$ such that for all $a \in M$, $a * e = e * a = a$.

Note: Since the requirements for a group contain the requirement for a monoid, every group is a monoid.

Example:

(a) The power set of any together with any one of the operations intersection, union, or symmetric difference is a monoid.

(b) The set of integers, **Z** with multiplication is a monoid. With addition, **Z** is also a monoid.

(c) The set of n × n matrices over the integers, M_n (**Z**), with matrix multiplication is a monoid. This follows from the fact that matrix multiplication is associative and has an identity, I_n. This is an example of a non-commutative monoid since there are matrices, A and B, for which AB and BA are different.

(d) $[\mathbf{Z}_n \times {}_n]$, n ≥ 2, is a monoid with identity 1.

(e) Let X be a non-empty set. The set of all functions from X into X, often denoted X^X, is a monoid over function composition, saw that function composition is associative. The function i: X → X defined by i (a) = a is the identity element for this system. This is another example of a non-commutative monoid, provided # X is greater than 1.

If X is finite, $\#(X^X) = \#X^{\#X}$. For example, if B = {0, 1}, $\#(B^B) = 4$. the functions z, u, i and t, defined by the graphs are the elements of B^B. This monoid is not a group. Do you know why? One reason that B^B is non-commutative is that tz ≠ zt, since (tz) (0) = 1 and (zt) (0) = 0.

FREE MONOIDS AND LANGUAGES

In this section, we will introduce the concept of a language. Languages are subsets of a certain type of monoid, the free monoid over an alphabet. After defining a free monoid, we will discuss languages and some of the basic problems relating to them. We will also discuss the common ways in which languages are defined.

Let A be a non-empty set, which we will call an *alphabet*. Our primary interest will be in the case where A is finite; however, A could be infinite for most of the situations that we will describe. The elements of A are called *letters or symbols*. Among the alphabets that we will use are B = {0, 1}, ASCII = the set of ADCII characters, and PAS = the Pascal character set (whichever one you use).

Example:

(1) English can be thought of as a language over the st of letters A, B, ... Z (upper and lower case) and other special symbols, such as punctuation marks and the blank. Exactly what subset of the strings over this alphabet defines the English language is difficult to pin

down exactly. This is a characteristic of natural languages that we try to avoid with formal languages.

(b) The set of all ASCII stream files can be defined in terms of a language over ADII. An ASCII stream file is a sequence of zero or more lines followed by an end-of-file symbol. A line is defined as a sequence of ADCII characters that ends with the two characters CR (carriage return) and LF (line feed). The end-of-file symbol is system-dependent; for example, CTRL/C is a common one.

(c) The set of all syntactically correct Pascal programs is a language over PAS. The set of Pascal text files is a language over PAS $\cup$ {eoln, eof}. The strings that belong to this language can be best described as the set of all strings over PAS $\cup$ {elon} concatenated with the eof marker. Recall that in Pascal, eoln and eof are not Char type values, but only markers. Despite this fact, they can still be considered part of the alphabet that defines text files.

(d) A few languages over B are L_1 = {s $\in$ B*; s has exactly as many 1's as it has 0's}, L_2 = {1s 0: s $\in$ B*} and L_3 = (0, 01) = the submonoid of B* generated by {0, 01}.

AUTOMATA, FINITE-STATE MACHINES

In this section, we will introduce the concept of an abstract machine. The machines that we will examine will (in theory) be capable of performing many of the tasks that are associated with digital computers. One such task is solving the recognition problem for a language. We will concentrate on one class of machines, finite-state machines (finite automata). And we will see that they are precisely the machines that are capable of recognizing strings in a regular grammar.

Given an alphabet X, we will imagine a string in X* to be encoded on a tape which we will call an *input tape.* When we refer to a tape, we might imagine a strip of material that is divided into segments, each of which can contain either a letter or a blank.

The typical abstract machine includes an input device, the *read head,* which is capable of reading the symbol from the segment of the input tape that is currently in the read head. Some more advanced machines have a read/write head that can also write symbols onto the tape. The movement of the input tape after reading a symbol depends on the machine. With a finite-state machine, the next segment of the input tape is always moved into the read head after a symbol has been read. Most machines (including finite-state machines) also have a separate *output tae* that is written n with a *write head.* The output symbols come from an output alphabet, Y, that may or may not be equal to the input alphabet. The most significant component of an abstract machine is its *memory structure.* This structure can range from a finite number of bits of memory (as in a finite-state machine) to an infinite amount of memory that can be store in the form of a tape that can be read from and written on (as in a Turing machine).

IDENTITY ELEMENT

Let S be a non-empty set and o be a binary operation on S.

(i) If there exists an element $e_1 \in S$, such that $e_1 oa = a$ for $a \in S$, then e_1 is called a left identity of S w.r.t. the operation.

(ii) If there exists an element $e_2, \in S$ such that $aoe_2 = a$ for $a \in S$, then e_2 is called a right identity of S w.r.t. the operation o.

(iii) If there exists an element $e \in S$ such that e is both a left and a right identity of S w.r.t. o, then e is called an identity of S.

Example:

(i) In the algebraic system (Z, +), the number 0 is an identity element.

(ii) In the algebraic system (R, .), the number 1, is an identity element.

(iii) Let (S, o) be an algebraic structure such that S contains atleast two elements and o be the operation such that aob = b for a, b $\in$ S. Then each element of S is a left identity of (S, o), but (S, o) has no right identity.

(iv) Let (S, o) be an algebraic structure such that S contains atleast two elements and o be the operation such that aob = a for a, b $\in$ S. Then each element of S is a right identity of (S, o) but (S, o) has no left identity.

(v) Let (S, .) be an algebraic structure such that S is the set of all even integers. It has neither a left identity nor a right identity.

Note : In an algebraic system (S, o).

(i) a left identity may exist and a right identity may not exist.

(ii) a right identity may exist and a left identity may not exist.

(iii) an identity may not exist.

Theorem 1(a):

Let (S, o) be an algebraic structure and if e is an identity of S w.r.t. o, then it is unique.

Proof:

If possible, let e, e' be identities of S.

$\therefore$ eoe' = e' (taking e as left identity)

and eoe' = e (taking e' as right identity)

$\therefore$ e' = e and hence identity in S with respect to o is unique.

This identity in S in called the identity element in S w.r.t. the operation o.

Note: If multiplication notation is used for a binary composition than the identity element w.r.t. the operation, if exists, is often denoted by 1, is called the multiplicative identity or unit element.

If additive notation is used then identity element, if exists, is often denoted by O, O is called the additive identity or zero element.

The elements 1 and O should not be confused with integers, although in special cases they may actually be integers.

Theorem 1(b):

Let (S, o) be an algebraic structure. If e_1 and e_2 be respectively left and right identities of S w.r.t. o, then $e_1 = e_2$.

Proof:

Since $e_1, e_2 \in S$ and e_1 is a left identity in S w.r.t. o, we have

$$e_1 \text{ o } e_2 = e_2 \qquad \text{...(1)}$$

Since $e_1, e_2 \in S$ and e_2 is a right identity in S w.r.t. o, we have

$$e_1 \text{ o } e_2 = e_1 \qquad \text{...(2)}$$

$\therefore$ From (1) and (2),

$$e_1 = e_2.$$

MONOID

A semi-group (S, o) with the identity element w.r.t. o is known as a monoid, i.e., (S, o) is a monoid if S is a non-empty set and o a binary operation in S such that o is associative and there exists an identity element w.r.t. o.

Example:

(i) (Z, +) is a monoid and the identity is O.

(ii) (Z, .) is a monoid and the identity element in 1.

(iii) S is the set of all mappings from a finite set A to itself and o is the composition of mapping in S. Then (S, o) is monoid with the identity element I (identity mapping).

(iv) Let S be the set of all 2×2 matrices such that each element is S are rational numbers. If matrix multiplication (.) is the binary operation on S, the (S, .) is a monoid and unit matrix I_2 is the identity element in S.

Similarly, if matrix addition (+) is the binary operation on S, then (S, +) is a monoid and null matrix O_2 is the identity element is S.

Note : If S is a finite set and o is a binary operation is S, we observe from the Cayley's composition table the following :

If the row (column) headed by an element a_i coincides with the first row (first column), then we say that identity exists in (S, o) and a_i is the identity in (S, o).

INVERTIBLE ELEMENT

Let (S, o) be an algebraic structure with the identity element in S w.r.t. o. An element a ∈ S in said to be left invertible or left regular if there exists an element x ∈ S such that xoa = e, x is called a left inverse of a w.r.t. o.

An element a ∈ S in said to be right invertible or right regular if there exists an element y ∈ S such that aoy = e, y is called a right inverse of w.r.t. o.

An element x which is both a left inverse and a right inverse of a is called an inverse of a and is said to be invertible or regular.

Theorem:

Let (S, o) be a monoid. If a ∈ S and a is invertible w.r.t. o, then inverse of a w.r.t. o is unique.

Proof:

Let e be the identity element of S w.r.t. o. Since a is invertible, it has an inverse w.r.t. o. If possible, let b ∈ S and c ∈ S be two inverses of a w.r.t. o in S.

$\therefore$ aob = e = boa and aoc = e = coa

Then co(aob) = coe = c ...(1)

and co(aob) = (coa)ob = eob = b ...(2)

$\therefore$ From (1) and (2), c = b

$\therefore$ Inverse of a is unique.

The unique inverse of a is denoted by a^{-1}, if the operation is taken multiplicatively and by – a if the operation is taken additively.

Note:

(i) The inverse of the identity element e is e.

[$\because$ eoe = e, i.e., $e^{-1} = e$]

(ii) $aoa^{-1} = a^{-1}oa = e$, $a + (-a) = (-a) + a = e$

(iii) $(a^{-1})^{-1} = a$, $-(-a) = a$, i.e.,

the inverse of the inverse of a is a.

CANCELLATION LAWS

Let S be non-empty set and o be a binary operation on S.

For a, b, c, ∈ S.

(i) $aob = aoc \Rightarrow b = c$,

(ii) $boa = coa \Rightarrow b = c$.

(i) is called *left cancellation law,*

(ii) is called *right cancellation law* and (i), (ii) are called *cancellation laws.*

SUB SEMI-GROUP AND SUBMONOID

Let (S, o) be a semi-group and T be a subset of S. If T is closed under the operation o, i.e., aob ∈ T whenever a and b are elements of T, then (T, o) is called a *subsemi-group* of (S, o).

Similarly, set (S, o) be a monoid with identity a, and let T be a non-empty subset of S. If T is closed under the operation o and e ∈ T, then (T, o) is called a *submonoid* of (S, o).

As associative property holds in any subset of a semi-group so that a subsemi-group (T, o) of a semi-group (S, o) is itself a semi-group. Similarly, a submonoid of a monoid is itself a monoid.

Examples:

(i) If T is the set of all even integers, then (T, .) is a subsemi-group of the monoid (Z, .), but it is not a submonoid since the identity of Z, the number 1, does not belong to T.

(ii) If (S, o) is a semi-group, then (S, o) is a subsemigroup of (S, o) Similarly, let (S, o) be a monoid. Then (S, o) is a submonoid of (S, o), and if T = {e}, then (T, o) is also a submonoid of (S, o).

RING

We have seen that the set I of all integers is an additive Abelian group with respect to the operation of ordinary addition. It is just as important to observe that I is also closed under ordinary multiplication and that multiplication is linked to addition in a way which enriches the structure of the system as a whole. The theory of rings is the theory of such systems.

A ring is an additive Abelian group R which is closed under a second operation called multiplication—the product of two elements x and y in R is written xy—in such a manner that :

(1) Multiplication is associative, that is, if x, y, z are any three elements in B, then $x(yz) = (xy)z$; and

(2) Multiplication is distributive, that is, if x, y, z are any three elements in R, then $x(y + z) = xy + xs$ and $(.x + y)z = xz + yz$.

In other words, a ring is an additive Abelian group whose elements can be multiplied as well as added, and in which multiplication behaves reasonably with respect to itself and addition. We note particularly that multiplication is not assumed to be commutative.

The concept of a group has its origin in the set of mappings or permutations, of a set onto itself. So far we have considered sets with one binary operation only. But rings are the outcome of the motivation which arises from the fact that integers follow a definite pattern with respect to the addition and multiplication. Thus we now aim at studying rings which are algebraic systems with two suitably restricted and related binary operations.

Definition : *An algebraic structure (**R**, + •), where **R** is a non-empty set and + and are . two defined operations in **R**, is called a ring if for all a, b, c, in **R**, the following axioms we are satisfied:*

R_1 . (R, +) is an abelian group, *i.e.*,

(R_{11}) $a + b \in R$ *(Closure law for addition)*

(R_{12}) $(a + b) + c = a + (b + c)$ *(Associative law for addition)*

(R_{13}) R has an identity, to be denoted by 0, with respect to addition,

i.e., $a + 0 = a \ \forall \ a \in R$ *(Existence of additive identity)*

(R_{14}) There exists an additive inverse for every element in R,

i.e.., there exists an element a in R such that

$a + (-a) = 0 \ \forall \ a \in R$ *(Existence of additive inverse)*

(R_{15}) $a + b = b + a$ *(Commutative law for addition)*

R_2 . (R,.) is a semigroup, *i.e.*,

(R_{21}) $a. b \in R$ *(Closure law/or multiplication)*

(R_{22}) $(a.b). c \Rightarrow a. (b. c)$ *(Associative law for multiplication)*

R_3 . Multiplication is left as well as right distributive over addition, *i.e.*,

$$a.(b + c) = a.b + a.c$$

and $$(b + c).a = b.a + c.a$$

ELEMENTARY PROPERTIES OF A RING

Theorem:

If **R** *is a ring, then for all a, b* $\in$ **R**.

(a) $a.0 = 0.a = 0$

(b) $a(-b) = (-a)b = -(ab)$

(c) $(-a)(-b) = ab.$

Proof :

(a) We know that

$a.0 = a(0 + 0) = a.0 + a.0 \; \forall \; a \in \mathbf{R}$ (Using distributive law)

Since **R** is a group under addition, applying right cancellation law,

$$a.0 = a.0 + a.0 \Rightarrow 0 + a.0 = a.0 + a.0 \Rightarrow a.0 = 0$$

Similarly, $0.a = (0 + 0)\, a = 0.a + 0.a$ (Using distributive law)

$$0 + 0.a = 0\, a + 0.a \qquad (\because 0 + 0.a = 0.a)$$

Applying right cancellation law for addition, we get

$$0 = 0.a \quad i.e., \; 0.a = 0$$

Thus, $a.0 = 0.a = 0.$

(b) To prove that $a(-b) = -ab$ we would show that

$$ab + a(-b) = 0$$

We know that $a\,[b + (-b)] = a.0 = 0$ $[\because b + (-b) = 0]$

(with the virtue of result (a) above)

or $ab + a(-b) = 0$ (by distributive law)

$\therefore$ $a(-b) = -(ab).$

Similarly, to show $(-a)\, b = -ab$, we must show that

$$ab + (-a)b = 0$$

But $ab + (-a)b = [a + (-a)]b = 0.b = 0$

$$-(a)b = -(ab)$$

Hence the result.

(c) Actually to prove $(-a)(-b) = ab$ is a special case of foregoing article. However its proof is given as under:

$(-a)(-b) = -[a(-b)]$ [by result b]

$= -[-(ab)]$ $[\because a(-b) = -ab]$

$= ab$

because $-(-x) = x$ is a consequence of the fact that in a group inverse of the inverse of an element is element itself.

Many of the additive Abelian groups listed in the previous section are also rings with respect to natural multiplications.

Example:

Each of the following rings consists of numbers, and addition and multiplication are understood to have their ordinary meanings.

(a) The single-element set containing only the number 0.

(6) The set I of all integers.

(c) The set of all even integers.

(d) The set of all rational numbers.

(e) The set R of all real numbers.

STRUCTURE OF RINGS

1. Commutative Rings

Definition : *A ring **R** is said to be commutative, if the multiplication composition in **R** is commutative. i.e.,*

$$ab = ba \ \forall \ a, b \in \mathbf{R}.$$

2. Rings with Unity Element

Definition : *A ring **R** is said to be a ring with unity element if **R** has a multiplicative identity, i.e., if there exists an element in **R** denoted by 1, such that*

$$a = a.1 = a \ \forall \ a \in R.$$

The ring of all n x n matrices with elements as integers (rational, real or complex numbers) is a ring with unity. The unity matrix

$$I_n = \begin{bmatrix} 1 & 0 & 0 & \dots\dots\dots & 0 \\ 0 & 1 & 0 & \dots\dots\dots & 0 \\ 0 & 0 & 1 & \dots\dots\dots & 0 \\ 0 & 0 & 0 & \dots\dots\dots & 1 \end{bmatrix}$$

is the unity element of the ring.

3. Rings with or without Zero Divisors

While dealing with an arbitrary ring R, we may find elements a and b in R neither of which is zero, and their product may be zero. We call such elements *divisors of zero or zero divisors.*

Definition : *A ring element a ($\neq 0$) is called a divisor of zero if there exists an element b ($\neq 0$) in the ring such that either*

$$ab = 0 \quad \text{or} \quad ba = 0$$

We also say that a *ring* R *is without zero divisors if the product of no two non-zero elements is zero. i.e, if*

$$ab = 0 \Rightarrow \text{either } a = 0 \quad \text{or } b = 0$$

or both $a = 0$ and $b = 0$.

THE MONOID OF A FINITE-STATE MACHINE

In this section, we will see how every finite-state machine has a monoid associated with it. For any finite-state machine, the elements of its associated monoid correspond to certain input sequences. Because only a finite number of combinations of states and inputs is possible for a finite-state machine, there is only a finite number of input sequences that summarize the machine. This idea is illustrated best with a few examples.

Example:

Consider the parity checker. The following table summarizes the effect on the parity checker of strings in B^1 and B^2. The row labelled "Even" contains the final state and final output as a result of each input string in B^1 and B^2 when the machine starts in the even state. Similarly, the row labeled "Odd" contains the same information for input sequences when the machine starts in the odd state.

Input String	0	1	00	01	10	11
Even	(Even, 0)	(Odd, 1)	(Even, 0)	(Odd, 1)	(Odd, 1)	(Even, 0)
Odd	(Odd, 1)	(Even, 1)	(Odd, 1)	(Even, 1)	(Even, 0)	(Odd, 1)
Same Effect as			0	1	1	0

Note how, as indicated in the last row, the strings in B^2 have the same effect as certain strings in B^1. For this reason, we can summarize the machine in terms of how it is affected by strings of length 1. The actual monoid that we will now describe consists of a set of functions, and the operation on the functions will be based on the concatenation operation.

Let T_0 be the final effect (state and output) on the parity checker of the input 0. Similarly, T_1 is defined as the final effect on the parity checker of the input 1. More precisely,

$$T_0 \text{ (even)} = \text{(even, 0) and } T_0 \text{ (odd)} = \text{(odd, 1)},$$

while

$$T_1 \text{ (even)} = \text{(odd, 1) and } T_1 \text{ (odd)} = \text{(even, 0)}$$

In general, we define the operation on a st of such functions as follows: if s, t are input sequences and T_s and T_t are functions as above, then $T_sT_t = T_{st}$, that is, the result of the function that summarizes the effect on the machine by the concatenation of s with t. Since, for example, 01 has the same effect on the parity checker as 1, $T_0T_1 = T_{01} = T_1$. We don't stop our calculation at T_{01} because we want to use the shortest string of inputs to describe the final result. A complete table for the monoid of the parity checker is

	T_0	T_1
T_0	T_0	T_1
T_1	T_1	T_0

What is the identity of this monoid? The monoid of the parity checker is isomorphic to the monoid $[\mathbf{Z}_2 + 2]$.

This operation may remind you of the composition operation on functions, but there are two principal differences. The domain of T_s is not the codomain of T_t, and the functions are read from left to right unlike in composition, where they are normally read from right to left.

You may have noticed that the output of the parity checker echoes the state of the machined and that we could have looked only at the effect on the machine as the final state. The following example has the same property, hence we will only consider the final state.

THE MACHINE OF A MONOID

Any finite monid [M; *] can be represented in the form of a finite-state machine with input and state sets equal to M. The output of the machine will be ignored here, since it would echo the current state of the machine. Machines of this type are called *state machines.* It can be shown that whatever can be done with a finite-state machine can be done with a state machine; however, there is a trade-off. Usually, state machines that perform a specific function are more complex than general finite-state machines.

Definition: *Machine of a Monoid. If [M; *] is a finite monoid, then the machine of M, denoted m(M), is the state machine with state set M, input set M, and next-state function t:* $M \times M \rightarrow$ *defined by* $t\,(s, x) = s * x$.

Example:

We will construct the machine of the monoid $[\mathbf{Z}_2: +_2]$. As mentioned above, the state set and the input set are both $\mathbf{Z}_2$. The next state function is defined by $t\,(s, x) = s +_2 x$. The transition diagram for m $(\mathbf{Z}_2)$ appears in note how it is identical to the transition diagram of the party checker, which has an associated monoid that was isomorphic to $[\mathbf{Z}_2; +_2]$.

CANCELLATION LAWS IN A RING

We say that cancellation laws hold in a ring R if

$$ab = ac \ (a \neq 0) \Rightarrow b = c$$

and $$ba = ca \ (a \neq 0) \Rightarrow b = c \text{ where a, b, c, are in R.}$$

Thus in a ring with zero divisors, it is impossible to define a cancellation law.

Theorem:

A ring has no divisor of zero if and only if the cancellation laws hold in R.

Proof:

Suppose that R has no zero divisors. Let a, b, c be any three elements of **R** such that $a \neq 0$, $ab = ac$.

Now, $ab = ac \Rightarrow ab - ac = 0$

$\Rightarrow \quad a(b - c) = 0$

$\Rightarrow \quad b - c = 0$ $\quad$ ($\because$ R is without zero divisors and $a \neq 0$)

$\Rightarrow \quad b = c.$

Thus the left cancellation law holds in R. Similarly, it can be shown that right cancellation law also holds in R.

Conversely, suppose that the cancellation laws hold in R.

Let $a, b \in R$ and if possible let $ab = 0$ with $a \neq 0$,

$b \neq 0$ then $ab = a.0$ $\quad$ ($\because a.0 = 0$)

Since $\quad a \neq 0, ab = a.0 \Rightarrow b = 0$ $\quad$ (by left cancellation law)

Hence, we get a contradiction to our assumption that $b \neq 0$ and therefore the theorem is established.

Division Ring : A ring is called a division ring if its non-zero elements form a group under multiplication.

Pseudo Ring : A non-empty set R with binary operations '+' and '.' satisfying all the postulates of a ring except right and left distributive laws, is called a pseudo ring if

$$(a + b).(c + d) = a.c + a.d + b.c + b.d \text{ for all } a, b, c, d \in R.$$

FUNDAMENTAL HOMOMORPHISM THEOREM

*Let $f : S \rightarrow T$ be a homomorphism of the semi-group (S, o) onto the semi-group (T, *). Let R be the relation on S defined by a R b if and only if $f(a) = f(b)$, for a and b in S. Then*

(a) R is a congruence relation.

*(b) (T, *) and the quotient semi-group (S/R, *) are isomorphic.*

Proof:

(a) First we will show that R is an equivalence relation.

(i) a R a for every $a \in S$, since f(a) = f(a). Hence R is reflexive

(ii) If a R b, then f(a) = f(b), so b R a. Hence R is symmetric.

(iii) If a R b and b R c, then f(a) = f(b) and f(b) = f(c), so f(a) = f(c) and a R c. So, R is transitive.

Hence R is an equivalence relation. Now suppose that a R a_1 and b R b_1. Then f(a) = f(a_1) and f(b) = f(b_1)

Multiplying in T, we get

$$f(a) * f(b) = f(a_1) * f(b_1).$$

Since f is a homomorphism, this equation can be written as

$$f(a \circ b) = f(a_1 \circ b_1)$$

Hence (a o b) R (a_1 o b_1)

and R is a congruence relation.

(b) We now consider the relation g from S/R to T defined as follows :

$$g = \{([a], f(a)) \mid [a] \in S/R\}.$$

We first show that g is a function. Suppose that [a] = [a']. Then a R a', so f(a) = f(a'), which implies that g is a function. We may now write $g : S/R \rightarrow T$, where

$$g([a]) = g(a) \text{ for } [a] \in S/R.$$

Now, we show that g is one to one. Suppose that g([a]) = g([a']). Then

$$f(a) = f(a').$$

So a R a', which implies that [a] = [a']. Hence g is one-one.

Now, we show that g is onto. Suppose that $b \in T$. Since f is onto, f(a) = b for some element a in S. Then

$$g([a]) = f(a) = b.$$

So g is onto.

Finally, $g([a] * [b]) = g([a \circ b])$

$= f(a \circ b) = f(a) * f(b)$

$= g([a]) * g([b]).$

Hence g is an isomorphism.

ISOMORPHISM AND HOMOMORPHISM

Isomorphism : *Let (S, o) and (T, *) be two semi-groups. A function f : S → T is called an isomorphism from (S, o) to (T, *) if it is a one-to-one correspondence from S to T, and if*

$f(aob) = f(a) * f(b)$

for all a and b in S.

If two semi-group (S, o) and (T, *) are isomorphic, we write $S \cong T$.

To show that two semi-groups (S, o), and (T, *) are isomorphic, we use the following procedure:

(i) Define a function f : S → T with Dom (f) = S.

(ii) Show that f is one-one.

(iii) Show that f is onto.

(v) Show that f(aob) = f(a) * f(b).

Theorem 1(a):

*Let (S, o) and (T, *) be monoids with identities e and e' respectively. Let f : S → T be an isomorphism. Then f(e) = e'.*

Proof:

Let b be any element of T. Since f is onto, there is an element a in S such that f(a) = b. Then

$$a = aoe$$

$$b = f(a) = f(aoe) = f(a) * f(e)$$

$$= b * f(e).$$

Similarly, since a = eoa, b = f(e) * b.

Thus for any $b \in T$,

$$b = b * f(e) = f(e) * b$$

which means that f(e) is an identity for T. Thus since the identity is unique, it follows that f(e) = e'.

Note: If (S, o) and (T, *) are semi-groups such that S has an identity and T does not, it then follows from theorem 2 that (S, o) and (T, *) can not be isomorphic.

Theorem 1(b):

*If f is an isomorphism from (S, o) to (T, *), then to show that f^{-1} is an isomorphism from (T, *) to (S, o).*

Proof:

Since f is one-one correspondence, then f^{-1} exists and is a one-one correspondence from T is S.

To show that f^{-1} is an isomorphism. Let a' and b' be any elements of T. Since f is onto, we can find elements a and b in such that f(a) = a' and f(b) = b'. Then a = f^{-1} (a') and b = f^{-1}(b'). Now $f^{-1}(a' * b') = f^{-1}(f(a) * f(b))$

$$= f^{-1}(f(aob))$$

$$= (f^{-1}of)\ (aob)$$

$$= aob = f^{-1}(a')\ of^{-1}(b')$$

Hence f^{-1} is an isomorphism.

HOMOMORPHISM

*Let (S, o) and (T, *) be two semi-groups. An everywhere defined function f : S → T is called a homomorphism from (S, o) to (T, *) if*

$$f(aob) = f(a) * f(b)$$

for all a and b in S. If f is also onto we say that T is a homomorphic image of S.

The difference between an isomorphism and a homomorphism is that an isomorphism must be one to one and onto. For both an isomorphism and a homomorphism, the image of a product is the product of the images.

Theorem 1:

*Let f be a homomorphism from a semi-group (S, o) to a semi-group (T, *). If S' is a subsemi-group of (S, o), then f(S') = {t ∈ T| t = f(s) for some s ∈ S'}, the image of S' under f, is a subsemi-group of (T, *).*

Proof:

If t_1 and t_2 are any elements of f(S'), then there exist s_1 and s_2 in S' with $t_1 = f(s_1)$ and $t_2 = f(s_2)$.

Then $$t_1 * t_2 = f(s_1) * f(s_2)$$

$$= f(s_1\ o\ s_2) = f(s_3)$$

where $s_3 = s_1 o s_2 \in S'$. Hence $t_1 * t_2 \in f(S')$.

Thus f(S') is closed under the operation *. Since the associative property holds in T, it holds in f(S'), so f(S') is a subsemi-group of (T, *).

Theorem 2:

*If f is a homomorphism from a commutative semi-group (S, o) onto a semi-group (T, *), then (T, *) is also commutative.*

Proof:

Let t_1 and t_2 be any elements of T. Then there exist s_1 and s_2 is S with $t_1 = f(s_1)$ and $t_2 = f(s_2)$.

Therefore,

$$t_1 * t_2 = f(s_1) * f(s_2) = f(s_1 \text{ o } s_2)$$
$$= f(s_2 \text{ o } s_1) = f(s_2) * f(s_1)$$
$$= t_2 * t_1$$

Hence (T, *) is also commutative.

SOLVED EXAMPLES

Example 1:

Show that the set M_n of all n X n matrices forms a ring with respect to addition and multiplication of matrices, when the elements of the matrices are real numbers.

Solution:

Let $A = [a_{ij}]$, $B = [b_{ij}]$ and $C = [c_{ij}]$ be any three elements of M_n, then we have

R_1 : (i) Since the sum of any two n×n matrices is defined and is an n x n matrix

i.e., $A + B = [a_{ij}] + [b_{ij}] = [a_{ij} + b_{ij}]$, say $\in M_n$

so the set M_n is closed with respect to addition of matrices.

(ii) Here $A + (B + C) = [a_{ij}] + ([b_{ij}] + [c_{ij}])$

$$= [a_{ij}] + [b_{ij} + c_{ij}] = [a_{ij} + (b_{ij} + c_{ij})]$$

i.e., $A + (B + C) = [(a_{ij} + b_{ij}) + c_{ij}]$, as addition of real numbers is associative.

$$= [a_{ij} + b_{ij}] + [c_{ij}]$$
$$= ([a_{ij}] + [b_{ij}]) + [c_{ij}] = (A + B) + C$$

i.e., the addition of matrices obeys associative law.

(iii) There exists the null matrix 0 of order n × n such that

$$0 + A = A = A + 0 \ \forall \ A \in M_n.$$

∴ The null matrix 0 is the additive unity.

(iv) There exists an unique matrix $-A = [-a_{ij}] \in M_n$ for each matrix $A = [a_{ij}] \in M_n$ such that

$$(-A) + A = 0 = A + (-A),$$

where 0 is the null matrix of order n x n.

i.e. additive inverse of each matrix $A \in M_n$ exists is $-A$.

(v) Here $A + B = [a_{ij}] + [b_{ij}] = [a_{ij} + b_{ij}]$

$= [b_{ij} + a_{ij}]$, as addition of real numbers is commutative

$= [b_{ij}] + [a_{ij}] = B + A$

i.e. the addition of matrices obeys commutative law.

R_2 : Since the product of any two $n \times n$ matrices is defined and is an $n \times n$ matrix, so the set M_n of matrices is closed with respect to multiplication of matrices.

R_3 : Since $A . (B . C) = (A . B) . C \ \forall A, B, C \in M_n$,

so the multiplication is associative.

$$R_4 : \text{since} \left.\begin{matrix} A . (B + C) = A . B + A . C \\ (B + C) . A = B . A + C . A \end{matrix}\right\} \forall A, B, C \in M_n$$

∴ The matrix multiplication is left and right distributive with respect to matrix addition.

Thus we find that all the postulates of ring are satisfied and hence the given set M_n of matrices is a ring with respect to addition and multiplication of matrices.

Example 2:

Fill in the blanks of the following composition table so that o is associative in S = {a, b, c, d}.

o	a	b	c	d
a	a	b	c	d
b	b	a	c	d
c	c	d	c	d
d				

Solution:

Using associative law and by trial and error :

Since do(aoa) = doa and (doa)oa

= do(aoa) = doa, doa must be equal to d only.

Sinced do(boa) = dob and (dob)

oa = do(boa), bob must be equal to a only.

Since $do(coa) = doc$ and $(doc)oa$

$= do(coa)$, doc must be equal to a only.

Since $do(doa) = dod$ and $do(doa)$

$= (dod)\ oa$, dod must be equal to a.

Thus d, a, a, a are respectively the four products

Example 3:

Let S be a non-empty set and o be an operation on S defined by aob = a for a, b $\in$ S. Determine whether o is commutative and associative in S.

Solution:

Since aob = a for a, b $\in$ S and boa = b for a, b $\in$ S.

aob $\neq$ boa.

$\therefore$ o is not commutative in S.

Since (aob)oc = aoc = a for a, b, c

ao(boc) = aob = a $\in$ S

$\therefore$ o is associative in S.

Examples 4(a):

A = {a, b}. Consider the set S of all mappings from A $\to$ A. Its the composition of mappings denoted by o is a binary composition in S.

Solution:

Total number of possible mappings from A $\to$ A is 4. Let these be :

$I : A \to A = \{(a, a), (b, b)\}$,

$f_1 : A \to A = \{(a, b), (b, a)\}$,

$f_2 : A \to A = \{(a, a), (b, a)\}$,

$f_3 : A \to A = \{(a, b), (b, b)\}$

$\therefore$ $S = \{I, f_1, f_2, f_3\}$. Let the composition of mappings be denoted by o.

Composition table is

o	I	f_1	f_2	f_3
I	I	f_1	f_2	f_3
f_1	f_1	I	f_2	f_3
f_2	f_2	f_3	f_3	f_3
f_3	f_3	f_2	f_2	f_3

Clearly,

(i) o is binary operation in S,

(ii) o is not commutative in S, and

(iii) o is associative in S.

Example 4(b):

Given that S = {A, B, C, D} where A = f, B = {a}, C = {a, b}. D = {a, b, c,}, show that S is closed under the binary operations ∪ (union of sets) and ∩ (intersection of sets) on S.

Solution:

$A \cup B = \phi \cup \{a\} = \{a\} = B.$

Similarly, $A \cup C = C, A \cup D = D$ and $A \cup A = A$

Also, $B \cup B = B, B \cup C = \{a\} \cup \{a, b\} = \{a, b\} = C,$

$B \cup D = \{a\} \cup \{a, b, c\} = \{a, b, c\} = D$

$C \cup C = C, C \cup D = \{a, b,\} \cup \{a, b, c\} = \{a, b, c\} = D.$

Hence ∪ is a binary operation on S.

Again, $A \cap A = A, A \cap B = \phi \cap \{a\} = \phi = A,$

$A \cap C = A, A \cap D = A$

and $B \cap B = B, B \cap C = \{a\} \cap \{a, b\} = \{a\} = B$

$B \cap D = \{a\} \cap \{a, b, c\} = \{a\} = B$

$C \cap C - C, C \cap D - \{a, b\} \cap \{a, b, c\} - \{a, b\} - C.$

Hence ∩ is a binary operation on S.

Example 4(c):

Show that the set of all integers .... –4, –3, –2, –1, 0, 1, 2, 3, 4 .... is an infinite abelian group with respect to the operation of addition of integers.

Solution:

Let us test all the group axioms for abelian group.

(G_1) Closure axiom : We know that the sum of any two integers is also an integer, *i.e.*, for all a, b ∈ I, a + b ∈ I. Thus I is *closed* with respect to addition.

(G_2) Associativity : Since the addition of integers is *associative*, the associative axiom is satisfied, *i.e.*, for a, b, c ∈ I.

$a + (b + c) = (a + b) + c$

(G_3) **Existence of Identity :** We know that 0 is the additive identity and $0 \in I$, *i.e.*,

$$0 + a = a = a + 0 \ \forall \ a \in I$$

Hence additive identity exists.

(G_4) **Existence of Inverse :** If $a \in I$, then $-a \in I$. Also,

$(-a) + a = 0 = a + (-a)$.

The every integer possesses additive inverse.

Therefore I is a *group* with respect to addition.

Since addition of integers is a commutative operation, therefore $a + b = b + a \ \forall \ a, \in I$.

Hence (I, +) is an abelian group. Also, I contains an infinite number of elements. Therefore (I, +) is an abelian group of infinite order.

Example 5:

Prove that the set **G** *= (0, 1, 2, 3, 4} is a finite abelian group of order 5 with respect to addition modulo 5.*

Solution:

Let us prepare a composition table as given below :

+ (mod 5)	0	1	2	3	4
0	0	1	2	3	4
1	1	2	3	4	0
2	2	3	4	0	1
3	3	4	0	1	2
4	4	0	1	2	3

Closure Property : All the entries in the composition table are elements of the set **G**. Hence **G** is closed under addition modulo 5.

Associative Property : Addition modulo 5 is associative always.

Identity : $0 \in$ **G** is the identity element.

Inverse. It is clear from composition table.

Element	–	0	1	2	3	4
Inverse	–	0	4	3	2	1

Inverse exists for every element of **G**.

Commutative law : The composition is commutative as the corresponding rows and columns in the table are identical.

The number of elements in **G** are 5.

Hence (**G**, + (mod 5)} is a finite abelian group of order 5.

Example 6:

Prove that the set **G** = *{1, 2, 3, 4, 5, 6} is a finite abelian group of order 5 with respect to multiplication modulo 7.*

Solution:

Let us prepare the following composition table :

$\times_7$	1	2	3	4	5	6
1	1	2	3	4	5	6
2	2	4	6	1	3	5
3	3	6	2	5	1	4
4	4	1	5	2	6	3
5	5	3	1	6	4	2
6	6	5	4	3	2	1

Closure Property : All the entries in the table are elements of G. Therefore G is closed with respect to multiplication modulo 7.

Associative Property : Multiplication modulo 7 is associative always.

Identity : Since first row of the table is identical to the row of elements of **G** in the horizontal border, the element to the left of first row in vertical border is identity element, *i.e.*, 1 is identity element in G with respect to multiplication modulo 7.

Inverse : From the table it is obvious that inverses of 1, 2, 3, 4, 5, 6 are 1, 4, 5, 2, 3 and 6 respectively. Hence inverse of each element in G exists.

Commutative Property : The composition is commutative because the elements equidistant from principal diagonal are equal each to each.

The set G has 6 elements. Hence (G, $\times_7$) is a finite abelian group of order 6.

Example 7(a):

Show that the set of all even integers (including zero) with additive property is an abelian group.

Solution:

The set of all even integers (including zero) is

$$I = \{0, \pm 2, \pm 4, \pm 6....)$$

Now we will discuss the group axioms one by one :

(G_1) The sum of two even integers is always an even integer, therefore closure axiom is satisfied.

(G_2) The addition is associative for even integers, hence associative axiom is satisfied.

(G_3) $0 \in I$, which is an additive identity in I, hence identity axiom is satisfied.

(G_4) Inverse of an even integer a is the even integer $-a$ in the set, so axiom of inverse is satisfied.

(G_5) Commutative law is also satisfied for addition of even integers. Hence the set forms an abelian group.

Example 7(b):

Show that the set of all non-zero rational numbers with respect to binary operation multiplication is a group.

Solution:

Let the given set be denoted by Q_0. Then by group axioms, we have

(G_1) We know that the product of two non-zero rational numbers is also a non-zero rational number. Therefore, Q_0 is closed with respect to multiplication.

Hence, closure axiom is satisfied.

(G_2) We know for rational numbers.

$$(a \,.\, b) \,.\, c = a \,.\, (b \,.\, c) \text{ for all } a, b, c \in Q_0$$

Hence, associative axiom is satisfied.

(G_3) Since, 1 the multiplicative identity is a rational number hence identity axiom is satisfied.

(G_4) If $a \in Q_0$, then obviously, $1/a \in Q_0$. Also

$$1/a \,.\, a = 1 = a \,.\, 1/a$$

so that 1/a is the multiplicative inverse of a. Thus inverse axiom is also satisfied.

Hence Q_0 is a group with respect to multiplication.

Example 8:

Let P(S) be the power set of a non-empty set S. Let '$\cap$' be an operation in P(S). Prove that associative law and commutative law are true for the operation $\cap$ in P(S).

Solution:

P(S) = set of all possible subsets of S.

Let $\quad A, B, C \in P(S)$

Since $\quad A \subset S, B \subset S \quad \Rightarrow A \cap B \subset S$

$\Rightarrow P \cap B \in P(S)$

Also $\quad B \cap A \subset S \quad \Rightarrow B \cap A \Rightarrow P(S).$

$\therefore \cap$ is a binary operation in P(S).

Also, $\quad A \cap B = B \cap A$

$\therefore \cap$ is commutative in P(S).

Again $A \cap B$, $B \cap C$, $(A \cap B) \cap C$ and $A \cap (B \cap C)$ are subsets of S.

$\therefore (A \cap B) \cap C, A \cap (B \cap C) \in P(S).$

Since $(A \cap B) \cap C = A \cap (B \cap C)$, $\cap$ is associative in P(S).

Example 9:

Show that C, the set of all non-zero complex numbers is a multiplicative group.

Solution:

Let $\mathbf{C} = \{z : z = x + iy, x, y \in \mathbf{R}\}$

Hence **R** is the set of all real numbers and $i = \sqrt{(-1)}$.

(G_1) Closure axiom : If $a + ib \in \mathbf{C}$ and $c + id \in \mathbf{C}$, then by definition of multiplication of complex numbers

$$(a + ib)(c + id) = (ac - bd) + i(ad + bc) \in C.$$

since $\quad ac - bd, ad + bc \in R$, for $a, b, c, d \in R$.

Therefore, C is closed under multiplication.

(G_2) Associative Axiom:

$(a + ib)\{(c + id) . (e + if)\} = (ace - adf - bcf - bdc) + i(acf + ade + bce - bdf) = \{(a + ib) . (c + id)\} . (e + if)$

for $a, b, c, d \in R$.

(G_3) Identity Axiom : $e = 1 (= 1 + i. 0)$ is the identity in **C**.

(G_4) Inverse Axiom : Let $(a + i b) (\neq 0) \in C$, then

$$(a+ib)^{-1} = \frac{1}{a+ib} = \frac{a-ib}{a^2+b^2}$$

$$= \left(\frac{a}{a^2+b^2}\right) + i\left(-\frac{b}{a^2+b^2}\right)$$

$$= m + i n \in \mathbf{C}, \text{where } m = \left(\frac{a}{a^2+b^2}\right), \; n = -\frac{b}{a^2+b^2} \in R$$

Hence **C** is a multiplicative group.

Example 10(a):

If S be the set of all real numbers of the form (m + n $\sqrt{2}$) where m, n ∈ Q, set of rational numbers, prove that S is a multiplicative or additive group, m, n not vanishing simultaneously.

Solution:

Let us first prove the question for multiplication as composition.

(G_1) Suppose $m_1 + n_1\sqrt{2}$, $m_2 + n_2\sqrt{2} \in S$,
so that $m_1, n_1, m_2, n_2 \in R$.
Now,

$(m_1 + n_1\sqrt{2}).(m_2 + n_2\sqrt{2}) = (m_1m_2 + 2n_1n_2) + (m_1n_2 + m_2n_1)\sqrt{2}$

Which belongs to S because $(m_1m_2 + 2n_1n_2), (m_1n_2 + m_2n_1) \in \mathbf{Q}$.

Hence *closure axiom* is satisfied.

(G_2) The elements of S are all real-numbers and the multiplication of real numbers is associative.

Hence *associative axiom* is satisfied.

(G_3) $1 = (1 + 0.\sqrt{2}) \in S$ is an identity element to multiplication.

(G_4) If $(m + n\sqrt{2}) \in S$ then

$$(m+n\sqrt{(2)}^{-1} = \frac{1}{m+n\sqrt{(2)}} = \frac{m-n\sqrt{2}}{m^2-2n^2}$$

$$= \left(\frac{m}{m^2-2n^2}\right) + \left(\frac{-n}{m^2-2n^2}\right)\sqrt{2}$$

$$= \alpha + \beta\sqrt{2} \in S$$

Where $\alpha = \frac{m}{m^2-2n^2}, \beta = \frac{-n}{m^2-2n^2} \in Q \; \forall \; m, n \in Q$.

Hence the *inverse axiom* is satisfied.

Therefore (S,.) is a group. Similarly, we can slow that (S, +) is also a group.

Example 10(b):

Show that the set of all n × n non-singular matrices having their elements as rational (real or complex) number is an infinite non-abelian group with respect to matrix multiplication.

Solution:

Let M be the set of all n × n non-singular matrices with their elements as rational numbers.

Let A ∈ M ⇒ A is a square matrix of order n, whose elements are rational numbers and |A| ≠ 0, because it is given that members of M are non-singular.

(G_1) Closure Property : If A, B ∈ M then AB will also be a matrix of order n × n, the elements of A will all be rational numbers, also |AB| = |A| |B| ≠ 0 as |A| ≠ 0 and |B| ≠ 0.

Thus AB ∈ M for A, B ∈ M. Therefore M is closed under matrix multiplication.

(G_2) Associativity : Multiplication of matrices is associative always.

(G_3) Existence of Identity : If I be the unit matrix of the type n × n, then the elements of I are all rational numbers. Also |I| = 1, *i.e.*, ≠ 0.

Therefore I ∈ M and also I A = A I ∀ A ∈ M.

Therefore I is identity element.

(G_4) Existence of Inverse : We known that inverse exist for all non-singular matrices. Therefore, if A ∈ M, there exists a non-singular matrix

$$A^{-1} = \frac{\text{Adj}A}{|A|}$$

with elements as rational numbers such that $A^{-1} A = I = AA^{-1}$.

Matrix multiplication, is not in general commutative. Number of elements in M is infinite.

Therefore M is an infinite non-abelian group with respect to matrix multiplication.

Example 10(c):

Prove that the set of matrices

$$A_\alpha = \begin{bmatrix} \text{Cos}\,\alpha & -\text{Sin}\,\alpha \\ \text{Sin}\,\alpha & -\text{Cos}\,\alpha \end{bmatrix}$$

where α is a real number, forms a group under multiplication.

Solution:

(G_1) Closure Property : Let $A_\alpha, A_\beta \in G$. Then

$$A_\alpha A_\beta = \begin{bmatrix} \text{Cos}\,\alpha & -\text{Sin}\,\alpha \\ \text{Sin}\,\alpha & \text{Cos}\,\alpha \end{bmatrix} \begin{bmatrix} \text{Cos}\,\beta & -\text{Sin}\,\beta \\ \text{Sin}\,\beta & \text{Cos}\,\beta \end{bmatrix}$$

$$= \begin{bmatrix} \text{Cos}\,(\alpha+\beta) & -\text{Sin}\,(\alpha+\beta) \\ \text{Sin}\,(\alpha+\beta) & \text{Cos}\,(\alpha+\beta) \end{bmatrix}$$

$$= A_{\alpha+\beta} \in G$$

(as $\alpha + \beta$ is a real number whenever α and β are real).

(G_2) Associativity. Let A_α, A_β and A_γ belong to G. Then

$$(A_\alpha \cdot A_\beta)A_\gamma = A_{\alpha+\beta} \cdot A_\gamma = A_{(\alpha+\beta)+\gamma}$$
$$= A_{\alpha+(\beta+\gamma)} = A_\alpha \cdot (A_\beta \cdot A_\gamma)$$

(G_3) Identity axiom : Since 0 is a real number, therefore

$$A_0 = \begin{bmatrix} \text{Cos}\,0 & -\text{Sin}\,0 \\ \text{Sin}\,0 & \text{Cos}\,0 \end{bmatrix} = \begin{bmatrix} 1 & 0 \\ 0 & 1 \end{bmatrix} \in G$$

Now, $A_\alpha \cdot A_0 = A_{\alpha+0} = A_\alpha$

Thus, $A_o = I_2$ is the identity element.

(G_4) Inverse Axiom : Let $A_\alpha \in G$, α is a real number. Then $A_{-\alpha} \in G$, since $-\alpha$ is also a real number, and we have

$$A_\alpha \cdot A_{-\alpha} = A_{(\alpha+(-\alpha))} = A_0 = A_{-\alpha} \cdot A_\alpha$$

Hence $A_{-\alpha}$ is the inverse of A_α.

Thus G is a group under matrix multiplication.

Example 11:

Prove that the set Q of all rational numbers other then 1 with the operation defined by a o b = a + b − ab constitutes an abelian group.

Solution:

(G_1) Closure Property : Let a, b $\in$ **Q** so that both a and b are rational numbers other than 1.

$\therefore$ a o b = a + b − ab is also a rational number other than 1 because if it is equal to 1,

i.e., if a + b − ab = 1 then a + b − ab − 1 = 0

or (a − 1)(b − 1) = 0.

This would mean that as = 1 or b = 1 which is not true. Hence aob ∈ Q, *i.e.*, the closure property is satisfied.

(G_2) Associative Law : Let a, b, c ∈ **Q**, then

$$(a \text{ o } b) \text{ o } c = (a + b - a b) \text{ o } c$$
$$= (a + b - ab) + c - (a + b - ab)c$$
$$= a + b + c - ab - ac - bc + abc.$$

Also, $$a \text{ o } (b \text{ o } c) = a \text{ o } (b + c - b c)$$
$$= a + (b + c - bc) - a (b + c - bc)$$
$$= a + b + c - ab - ac - bc + abc.$$

Thus (a o b) o c s = a o (b o c) and hence associative law is satisfied.

(G_3) Existence of Identity : If e be the identity then

$$a \text{ o } e = a \Rightarrow a + e - ae = a$$

$\Rightarrow$ e (l – a) = 0 e = 0 as a ≠ 1

and in the case a o e = a o 0 = a + 0 – a. 0 = a

Hence 0 is the identity.

(G_4) Existence of Inverse :

a o b = a + b – ab = 0 the identity.

$\Rightarrow \quad \dfrac{a}{a-1} = b$ is the inverse of a

Also, since a – 1 ≠ 0

∴ a/a – 1 is a rational number other than 1 and as such belongs to **Q**,

Therefore inverse exists and belongs to the set.

(G_5) Commutative Law:

$$a \text{ o } b = a + b - ab = b + a - ba = b \text{ o } a$$

Hence commutative.

Therefore, Q is an abelian group.

Example 12:

I is the set of all integers (positive, negative and zero); is the ordinary multiplication and + is the ordinary addition of integers. Then show that I is a commutative ring with unit element.

Solution:

For the set I we find that:

(i) *Closure Property.* We know the sum and product of two integers is also an integer *i.e.* $a + b \in I$, $a \cdot b \in I \ \forall\ a, b \in I$

Thus the set of integers is closed with respect to addition and multiplication.

(ii) $a + b = b + a \ \forall\ a, b \in I$

i.e. addition of integers is commutative.

(iii) *Associative Laws for addition and multiplication.*

$$(a + b) + c = a + (b + c) \text{ and } (a \cdot b) \cdot c = a \cdot (b \cdot c) \ \forall\ a, b, c \in I.$$

i.e. the addition and multiplication of integers is associative.

(iv) $a \cdot (b + c) = a \cdot b + a \cdot c$ and $(b + c), a = b \cdot a + c \cdot a \ \forall\ a, b, c \in I.$

i.e. the multiplication is both left and right distributive with respect to addition.

(v) There exists an element 0 in I, such that

$$a + 0 = 0 + a = a \ \forall\ a \in I$$

(vi) There exists an element – a in I such that

$$a + (-a) = 0 = (-a) + a, \ \forall\ a \in I$$

Thus all the postulates are satisfied and hence (I, +,.) is a ring.

Also $\quad a \cdot b = b \cdot a \ \forall\ a \in I$

∴ I is a commutative ring.

Also $\quad 1 \cdot a = a \cdot 1 = a \ \forall\ a \in I$

∴ (I, + ,.) is a ring with unit element.

Example 13:

Show that a finite set **G** *with an associative binary operation is a group if the right and left cancellation laws hold in* **G**. *Show further that validity of the cancellation laws does not characterize infinite groups but characterize finite groups only.*

Solution:

Let G = $\{a_1, a_2, a_3 ..., a_n\}$ where $a_1, a_2, ... a_n$ are distinct. Let us consider an element a of G then $a_1a, a_2a, a_2a,... a_n a \in G$ the set G being closed under the binary operation and these elements are all distinct, because, otherwise $a_ia = a_ja$, for $a_i, a_j \in G$ $(i \neq j)$ for $a_i = a_j$ by the right cancellation law.

Thus $a_1a, a_2a, a_3a, ..., a_n a$ are nothing but some permutation of elements $a_1, a_2, a_3,...a_n$ of G.

Therefore, if b is an element of G, then it must be one of the elements $a_1a, a_2a,..... a_na$.

Hence $a_i a = b$, where a, b, $a_i \in G$.

Thus we arrive at the conclusion that the equation ya = b(a, b $\in$ G) has a unique solution in G. Similarly by forming the products aa_1, aa_2, aa_3, ... aa_n and by using the left cancellation law,

We can show that the equation ax = b (a, b $\in$ G) has a solution in G.

Hence G is a finite group of order n.

However, an infinite set will not necessarily form a group even if the binary composition in a set is associative and both the cancellation laws hold good in it. The following example will illustrate this fact.

The set N of all natural numbers 1, 2, 3, 4, ... is not a group for multiplication even though N is closed under multiplication, multiplication is associative and both the cancellation laws hold good for multiplication in N.

Example 14(a):

Show that the set N of natural numbers is not a ring with respect to addition and multiplication.

Solution:

If (N, +,.) is a ring, then (N, +) must form a group.

But we find that there exists no identity element e in N such that

$$a + e = a = e + a \ \forall \ a \in N$$

Hence the set N of natural numbers is not an additive group and consequently the set N is not a ring.

Example 14(b):

Show that the set of vectors defined as directed line segments does not form a group (i) with respect to secular (dot) product (ii) with respect to vector (cross) product.

Solution:

Let V denote the set of all vectors defined as directed line segments.

1. If a, b $\in$ V, then a. b is a scalar quantity and so a. b $\notin$ V. Thus dot product of vectors is not a binary operation on the set V. Hence V cannot be a group with respect to dot product.
2. If a, b $\in$ V, then the cross product a $\times$ b is also a vector and so a $\times$ b $\in$ V. Thus, cross product of two vectors is a binary operation on the set V. But the cross product of vectors is not an associative operation. If a, b, c $\in$ V, then in general

$a \times (b \times c) \neq (a \times b) \times c.$

Hence V is not a group with respect to cross product.

Example 14(c):

Show that the following are groups:

1. *The set C of all complex numbers with respect to the operation of addition of complex numbers.*
2. *The set Q of all rational numbers with respect to addition.*
3. *The set R of all real numbers with respect to addition.*

Solution:

1. Do your self.

(**Hint:** Here if $x_1 + iy_1$ and $x_2 + iy_2$ are any two complex numbers, then $(x_1 + iy_1) + (x_2 + iy_2) = (x_1 + x_2) + i(y_1 + y_2)$ is also a complex number. Thus, C is closed for addition of complex numbers or in other words addition is a binary operation on the set of complex numbers. The addition of complex numbers is an associative operation. The zero complex number *i.e.*, the complex number $0 + i0$ is identity for addition of complex numbers. Finally if $x + iy$ is any complex number, then the complex number $-x - iy$ its additive inverse. Thus, C is a group for addition of complex numbers).

2. We have $Q = \left\{\frac{a}{b} : a, b \in I \text{ and } b \neq 0\right\}$,

   Do your self.

   Hint: Here the rational number 0 is the additive identity and if $x = a/b$ is any rational number, then the rational number $-a/b$ is its additive inverse.

3. Do your self.

   Hint: Here the real number 0 is the additive identity and if a is any real number, then the real number $-a$ is its additive inverse.

Example 15:

Show that the set N of all natural numbers 1, 2, 3, 4, 5,... does not form a group with addition or multiplication but it forms a semi-group with respect to addition as well as multiplication.

Solution:

1. **Closure Property:** We know that the set N of natural numbers is closed with respect to addition as well as with respect to multiplication. Therefore both addition and multiplication are binary operations on the set N.

2. **Association Property:** We know that also both addition and multiplication of natural numbers are associative operations.

   Hence both the algebraic structures (N, +) and (N, .) are semi-groups.

3. **Identify Element:** Hence, the algebraic structure (N, +) is not a group because it does not possess identity element. There exists no natural number $e \in N$ such that $e + a = a = a + e \ \forall \ a \in N$. For the addition of numbers, the number 0 is the identify and $0 \notin N$. Therefore (N, +) is not a group.

Again the algebraic structure (N, .) is also not a group. This algebraic structure possesses identity element and it is the natural number 1. We have $1a = a = a1 \ \forall \ a \in N$. But except 1 no other natural number possesses multiplicative inverse. Hence (N, .) is not a group.

Example 16:

*Show that the set of positive rational numbers does not form a group with respect to the binary operation * defined by*

$$a * b = a/b.$$

Solution:

The set Q_+ of positive rational numbers is closed for the operation *. If a, $b \in Q_+$, then $a * b = a/b$ is also a positive rational number and so $a + b \in Q_+$.

But the operation on Q_+ defined by * is not associative. If a, b, $c, \in Q_+$, then in general $(a/b)\ c \neq a/(b/c)$.

Hence (Q_+, *) is not a group.

Important Note: In the few examples of groups we have just given, the elements of the underlying sets were all numbers. But its should not be confused that we can have groups of numbers only. In our definition of a group, we have not imposed any restriction on the elements of the set G. The elements of G can be numbers, matrices, chairs, students, countries or any thing. The set G will be a group if it is equipped with a binary composition satisfying certain postulates. Now we shall prove certain general properties of groups. While proving these properties we should not forget that G is any set.

Example 17(a):

Show that the set I of all integers

$$\{..., -4, -3, -2, -1, 0, 1, 2, 3, 4, ...\}$$

is a group with respect to the operation of addition of integers.

Solution:

1. **Closure Property:** We know that the sum of two integers is also an integer *i.e.*, $a + b \in \mathbf{I}\ \forall\ a, b \in \mathbf{I}$. Hence **I** is closed with respect to addition.
2. **Associativity:** We know that associative property w.r.t. addition.
3. **Existence of Identity:** Then number $0 \in I$. Also we have $0 + a = a = a + 0\ \forall\ a \in \mathbf{I}$. Therefore the integer 0 is the identity.
4. **Existence of Inverse:** If $a \in \mathbf{I}$, then $-a \in \mathbf{I}$. Also we have $(-a) + a = 0 = a + (-a)$. Thus every integer possesses additive inverse.

Therefore **I** is a group with respect to addition. Since addition of integers is commutative composition, therefore (**I**, +) is an abelian group. Also **I** contains an infinite number of elements. Therefore (**I**, +) is an abelian group of infinite order.

Example 17(b):

Do the positive irrationals form a group with respect to multiplication?

Solution:

The set of positive irrationals is not closed for multiplication. For example, $\sqrt{2}. \sqrt{2} = 2$ which is a rational number and so it does not belong to the set of positive irrationals.

Example 18:

Prove that the set Q_0 of all non-zero rational numbers forms a group under the operation of multiplication of rational numbers.

Solution:

Closure Property: We know that the product of two non-zero rational numbers is also a non-zero rational number, Therefore Q_0 is closed with respect to multiplication.

Associativity: We know that multiplication of rational numbers is an associative composition.

Existence of Identity: The rational number $1 \in Q_0$. Also we have

$$1a = a = a1\ \forall\ a \in Q_0.$$

The rational number 1 is the multiplicative identity.

Existence of Inverse: If $a \in Q_0$, then obviously $1/a \in Q_0$. Also $(1/a)\,a = 1 = a\,(1/a)$. Therefore 1/a is a multiplicative inverse of a.

Hence Q_0 is a group with respect to multiplication.

Since ab = ba $\forall$ a, b, $\in Q_0$, therefore the group is abelian.

Note: The set Q of all rational numbers is not a group with respect to multiplication. The rational number $0 \in Q$, but $\exists$ no rational number $a \in Q$ such that $0a = 1$.

We know that $0a = 0 \;\forall\; a \in Q$. Thus, the rational number 0 does not posses multiplicative inverse.

Example 19:

Does the set of all odd integers form a group with respect to addition

Solution:

We know that the sum of two odd integers is an even integer. For example, 3 + 7 = 10 which is even. Therefore, the set of odd integers is not closed with respect to addition. Thus, addition is not a binary composition in the set of odd integers. Therefore the question of the set of odd integers becoming a group with respect to addition does not arise.

Example 20:

Show that the set of all positive rational numbers forms an abelian group under the composition defined by

$$a * b = (ab)/2.$$

Solution:

We Q_+ denote the set of all positive rational numbers by Q_+. Now we show that $\{Q_+, *\}$ is a group:

Closure Property: Since for very a, b, $\in Q_+$, (ab)/2 is also in Q_+, therefore Q_+ is closed with respect to the operation *.

Associativity. Let a, b, c, $\in Q_+$. Then

$$(a * b) * c = \left(\frac{ab}{2}\right) * c = \frac{[(ab)/2]c}{2} = \frac{a[(bc)/2]}{2} = a * \left(\frac{bc}{2}\right) = a * (b * c)$$

Commutativity: Let a, b, $\in Q_+$. Then a*b = (ab)/2 = (ba)/2 = b*a.

Existence of Identity: Then number e will be the identity element if $e \in Q_+$ and if $e * a = a = a * e \;\forall\; a \in Q_+$.

Now $e * a = a \Rightarrow (ea)/2 = a$

$\Rightarrow (a/2)\,(e - 2) = 0$

$\Rightarrow e = 2$, since $a \in Q_+ \Rightarrow a \neq 0$.

Now $2 \in Q_+$ and we have $2*a = (2a)/2 = a = a * 2 \;\forall\; a \in Q_+$.

$\therefore$ 2 is the identity element.

Existence of Inverse: Let a be any element of Q_+. If the number b is to be the inverse of a, then we must have

$$b * a = e = 2 \Rightarrow (ba)/2 = 2 \Rightarrow b = 4/a.$$

Now $\quad a \in Q_+ \Rightarrow 4/a \in Q_+$.

We have $(4/a) * a = \{(4/a)\, a\}/2 = 2 = a * (4/a)$.

Therefore 4/a is the inverse of a. Thus each element of Q_+ is inversible.

Hence $(Q_+, *)$ is an abelian group.

Example 21(a):

Show that the set

$$G = \{..., -4m, -3m, -2m, -m, 0, m, 2m, 3m, 4m, ...\}$$

of multiples of integers by a fixed integer m, is a group with respect to addition.

Solution:

Closure Property: Let a, b be any tow elements of G. Then a = rm and b sm where r and s are some integers.

Now a + b = rm + sm = (r + s) m. Since r + s is also an integer, therefore (r + s) m *i.e.*, $a + b \in G$. Thus, $a + b \in G\ \forall\ a, b \in G$. Therefore G is closed with respect to addition.

Associativity: We know that the addition of integers is an associative composition and elements of G are all integers.

Existence of Identity: $0 \in G$ and we have $0 + a = a = a + 0\ \forall\ a \in G$. Therefore 0 is the identity.

Existence of Inverse: Let rm be an arbitrary element of G where r is some integer.

Then $\quad (-r)\, m \in G. \qquad [\because\ -r$ is also an integer]

Also $\quad (-r)\, m + rm = (-r + r)\, m = 0m = 0$

and $\quad rm + (-r)\, m = (r - r)\, m = 0.$

$\therefore$ (–r) m is the additive inverse of rm.

Thus every element of G possesses additive inverse.

Hence G is a group with respect to addition.

Example 21(b):

Is the set of integers

$$..., -3, -2, -1, 0, 1, 2, 3, ...$$

a group (1) with respect to subtraction (2) with respect to multiplication.

Solution:

1. We have $a - b \in \mathbf{I} \ \forall \ a, b, \in \mathbf{I}$. Therefore **I** is closed with respect to subtraction. But subtraction in **I** is not an associative composition. For example,

   $$(7 - 3) - 2 = 4 - 2 = 2,$$
   $$7 - (3 - 2) = 7 - 1 = 6.$$

   Thus $(7 - 3) - 2 \neq 7 - (3 - 2)$. Therefore (I,–) is not a group.

2. We have $ab \in I \ \forall \ a, b \in I$. Therefore I is closed with respect to multiplication. Also multiplication of integers is an associative composition. The integer 1 is the multiplicative identity since $1a = a = a1 \ \forall \ a \in I$. But 1 and –1 are the only integers which possess multiplicative inverse. Inverse of 1 is and the inverse of –1 is –1. However $2 \in I$ and the inverse of 2 for multiplication would have been 1/2 which is not an element of I.

   Therefore I is not a group with respect to multiplication.

Example 21(c):

Show that the set of all even integers with zero is an abelian group with respect to addition.

Solution:

Proceed as in example (before) taking $m = 2$. Here the sum of two even integers is also an even integer. The even integer 0 is additive identity. If r is any even integer, then –r is also an even integer and is the additive inverse of r.

Example 22:

Prove that the set of all rational numbers is a ring with respect to ordinary addition and multiplication.

Solution:

Let Q be the set of all rational numbers.

$\mathbf{R}_1$. (Q, +) is abelian.

(R_{11}) Let $a, b \in Q$ then $a + b \in Q$ because sum of two rational numbers is a rational number.

(R_{12}) Let $a, b, c \in Q$ then

$(a + b) + c = a + (b + c)$

because associative law for addition holds.

(R_{13}) $0 \in Q$ and $0 + a = a + 0 = a \ \forall \ a \in Q$, *i.e.*, 0 is the additive identity in Q.

(R_{14}) $\forall \ a \in Q, - a \in Q$ and $a + (-a) = 0$. Hence additive inverse in Q exists for each element in Q.

(R_{15}) Let $a, b \in Q$ then $a + b = b + a$ because addition is commutative for rationals.

R_2 (Q, •) is a semi-group.

(R_{21}) Since the product of two rational numbers is a rational number,

$$a, b \in Q \Rightarrow a \, . \, b \in Q.$$

(R_{22}) Multiplication in Q is associative.

R_3 Multiplication is left as well as right distributive over addition in the set of rational numbers, *i.e.*,

$$a \, . \, (b + c) = a \, . \, b + a \, . \, c$$

$$(b + c) \, . \, a = b \, . \, a + c \, . \, a,$$

$$\text{for } a, b, c, \in Q.$$

Hence (Q, +,.) is a ring.

Example 23:

Prove that the set E of even integers with zero, is a ring with respect to ordinary addition and multiplication.

Solution:

Here the set $E = \{0, \pm 2, \pm 4, \pm 6, ...\}$

For this set E we find that

R_1 : (i) $a + b \in E, \forall \ a, b \in E$

i. e. the set of even integers is closed with respect to addition.

(ii) $(a + b) + c = a + (b + c), \forall \ a, b, c \in E$

i.e. the associative law for addition holds in E.

(iii) There exists an identity element $0 \in E$, such that

$$a + 0 = 0 + a = a, \forall \ a \in E$$

(iv) For each $a \in E$, there exists an element $-a \in E$, such that

$$a + (-a) = 0 = (-a) + a$$

(v) $a + b = b + a, \forall \ a, b \in E$

i.e. the commutative law holds for addition in E.

$R_2 : a . b \in E, \forall a, b \in E$

i.e. the set E is closed with respect to multiplication.

$R_3 : (a . b) . c = a . (b . c), \forall \in E.$

i.e. the multiplication composition is associative in E.

$R_4 : a . (b + c) = a . b + a . c$

and $(b + c) . a = b . a + c . a, \forall a, b, c \in E$

i.e. the multiplication is left and right distributive over addition in E.

Thus all the ring axioms are verified and hence E is a ring with respect to ordinary addition and multiplication.

Example 24:

*Show that if every element of a group **G** is its own inverse, then **G** is abelian.*

Solution:

Let a and b be any two elements of G. Then ab is also an element of G.

Therefore, $(ab)^{-1} = ab$ (because it is given that every element in G is its own inverse)

Hence, $(ab)^{-1} = ab \Rightarrow b^{-1} a^{-1} = ba$

$\Rightarrow$ $ba = ab$ $[\because a^{-1} = a, b^{-1} = b]$

Hence, we have $ab = ba \ \forall a, b \in G$

$\therefore$ G is an abelian group.

Example 25(a):

Prove that the set of all real numbers of the form $m + n\sqrt{2}$ where m, n are rational numbers is a ring under the usual addition and multiplication.

Solution:

Let $R = \{m + n\sqrt{2} : m, n \text{ are real numbers}\}$.

$\mathbf{R_1}$ (R, +) is abelian group.

(R_{11}) Let $m_1 + n_1\sqrt{2}$, $m_2 + n_2\sqrt{2} \in R$, then

$(m_1 + n_1\sqrt{2}) + (m_2 + n_2\sqrt{2}) = (m_1 + m_2) + (n_1 + n_2)\sqrt{2} \in R$

because sum of two real numbers is a real number.

(R_{12}) $(m_1 + n_1\sqrt{2}) + (m_2 + n_2\sqrt{2}) = (m_2 + n_2\sqrt{2}) + (m_1 + n_1\sqrt{2})$

because addition of real numbers is a real number.

(R_{13}) Associative law for addition of real numbers holds, *i.e.*,

$$(m_1 + n_1 \sqrt{2}) + \{(m_2 + n_2 \sqrt{2}) + (m_3 + n_3 \sqrt{2})\}$$
$$= \{(m_1 + n_1 \sqrt{2}) + (m_2 + n_2 \sqrt{2})\} + (m_3 + n_3 \sqrt{2})$$

for $m_1, n_1, m_2, n_2, m_3, n_3$, to be rational numbers.

(R_{14}) $0 \ (= 0 + 0 \,.\, \sqrt{2}) \in R$ is the identity of addition in R.

(R_{15}) Let $m + n\sqrt{2} \in R$, then $-(m + n\sqrt{2})$

$$= -m - n\sqrt{2} \in R \text{ and also}$$

$$(m + n\sqrt{2}) + (-m - n\sqrt{2}) = (m - m) + (n - n)\sqrt{2} = 0$$

Hence additive inverse for each element in R exists in R.

R_2 (R,.) is a semi-group.

(R_{21}) $(m_1 + n_1 \sqrt{2}) \,.\, (m_2 + n_2 \sqrt{2})$

$$= (m_1 m_2 + 2n_1 n_2) + (m_1 n_2 + m_2 n_1)\sqrt{2}$$
$$= a + b\sqrt{2} \in R$$

as a and b being the sums of products of rational numbers are rational.,

(R_{22}) Multiplication is associative in R, *i.e.*,

$$\{(m_1 + n_1 \sqrt{2}) \,.\, (m_2 + n_2 \sqrt{2})\} \,.\, (m_3 + n_3 \sqrt{2})$$
$$= (m_1 + n_1 \sqrt{2}) \,.\, \{(m_2 + n_2 \sqrt{2}) \,.\, (m_3 + n_3 \sqrt{2})$$

R_3 Multiplication is left as well as right distributive over addition in R.

Hence R is a ring under usual addition and multiplication.

Example 25(b):

Prove that the set of integers $S = [m\lambda : \lambda, \in Z]$, where Z is the set of all integers and m is a fixed integer with addition and multiplication of integers as compositions is a commutative ring.

Or

Prove that the set $S = \{\lambda m\}$, where m is a fixed integer and $\lambda = 0, \pm 1, \pm 2, \pm 3,$ is a commutative ring with respect to ordinary addition and multiplication of integers. Is it a ring with unity ?

Solution:

If $\lambda_1 \lambda_2, \lambda_3$, be any three integers out of $0, \pm 1, \pm 2, \pm 3, \ldots$, then we have:

R_1 : (i) $\lambda_1 m \in S, \lambda_2 m \in S \Rightarrow (\lambda_1 m + \lambda_2 m) = (\lambda_1 + \lambda_2)\, m \in S$, since $\lambda_1 + \lambda_2$ is also an integer out of $0, \pm 1, \pm 2, \ldots$

i.e. the set S is closed with respect to addition.

(ii) We know that the integers obey the associative law of addition, so the set S obeys the associative law of addition as its elements are also integers.

(iii) There exists the additive identity 0m or 0 such that

$$\lambda m + 0m = \lambda m = 0m + \lambda m \ \forall \ \lambda m \ \forall \ S$$

(iv) For each λ in m S there exists an element $(-\lambda)$ m $\in$ S such that $\lambda m + (-\lambda) m = [\lambda + (-\lambda)]m = 0m$, the identity element and

$$(-\lambda) m + \lambda m = [(-\lambda) + \lambda] m = 0m$$

i.e. the additive inverse of each element of S exists.

(v) $\lambda_1 m + \lambda_2 m = (\lambda_1 + \lambda_2) m = (\lambda_2 + \lambda_1) m$, since commutative law holds for integers $= \lambda_2 m + \lambda_1 m$

i.e. the commutative law holds for addition in S.

R_2 : $(\lambda_1 m) . (\lambda_2 m) = (\lambda_1 \lambda_2) m \in S$, as $\lambda_1 \lambda_2$ is also an integer.

i.e. the set S is closed with respect to multiplication.

R_3 : We know that integers obey the associative law of multiplication so the set S obeys the associative law of multiplication, since its elements are integers.

R_4 : Since the integers are left and right distributive with respect to addition, so the multiplication is left and right distributive over addition in S.

R_5 : Since the multiplication is commutative in the set of integers, so the commutative law of multiplication holds in S, whose elements are integers.

Hence all the postulates for the ring are satisfied and so (S, +,.) is a ring.

Moreover if $m \neq 1$, then $1 \notin S$ and so it is a ring without unity. But if $m = 1$, then S is a set of integers and $1 \in S$ and so it is a ring with unity.

Example 25(c):

Prove that the set R of residue classes modulo m, where m is a positive integer, is a ring with respect to addition and multiplication of residue classes (mod m).

Solution:

Here we have

$$R = [\{0), \{1\}, \{2\}, ..., \{p\}, ... , \{q\}, ... \{m - \}]$$

R_1 : (i) Since $\{p\} + \{q\} = \{p + q\} \in R$,

where p + q is reduced (mod m) if $p + q \geq m$.

For example $5 + 8 = 13 = 3$ (mod. 10)

So the set R is closed with respect to addition.

(ii) Since the addition of integers is associative,

So $\forall$ {p}, {q}, {r} $\in$ R we have

$$\{p\} + [\{p) + \{r\}] = \{p\} + \{q + r\}$$
$$= \{p + (q + r)] = [(p + q) + r\}$$
$$= \{p + q\} + \{r\} + [\{p\} + \{q\}] + \{r\}$$

i.e. the addition of residue classes mod. m is associative.

(iii) There exists the additive identity [p], since

$$\{p\} + \{0\} = \{p\} = \{0\} + \{p\} \ \forall \ \{p\} \in R$$

(iv) For each {p} $\in$ R, there exists {–p} $\in$ R, such that

$$\{p\} + \{-p\} = \{p + (-p)\} = \{0\}$$

(v) Since the addition of integers is commutative, so we have

$$\{p\} + \{q\} = \{p + q\} = \{q + p\} = \{q\} + \{p\}$$

i.e. the addition in R is commutative.

Hence the residue classes modulo m form an abelian group with respect to addition of residue classes.

$\mathbf{R_2}$: Since [p] . [q] = {pq} $\in$ R

where pq is reduced (mod m) if $pq \geq m$

For example $5.8 = 40 = 0$ (mod 10) or 4 (mod 9)

Hence the set R is closed under multiplication of residue classes mod m.

$\mathbf{R_3}$: Since multiplication of integers is associative, so

$\forall$ [p], {q}, {r} $\in$ R we have

$$\{p\} . [\{q\} . \{r\}] = \{p\} . \{qr\} = \{p \{qr\}]$$
$$= \{(pq) r\} = \{pq\} . \{r\}$$
$$= [\{p\}. \{q\}] . \{r\}$$

i.e. the multiplication of residue classes mod m is associative.

$\mathbf{R_4}$: $\forall$ {p}, {q} and {r} $\in$ R, we have

$$\{p\} . [\{q\} + \{r\}] = \{p\} . \{q + r\}$$
$$= \{p . (q + r)] = \{pq + pr]$$
$$= \{pq\} + \{pr\}$$
$$= \{p\} . \{q) + \{p\} . \{r\}$$

Similarly we can show that

$$[\{q\} + \{r\}] \cdot \{p\} = \{q\} \cdot \{p\} + \{r\} \cdot \{p\}$$

i.e. multiplication of residue classes mod m is left and right distributive over addition of residue classes mod m.

R_5 : Since the multiplication of integers is commutative, so

$\forall$ {p}, {q} $\in$ R we have

$$\{p\} \cdot \{q\} = \{pq\} = \{qp\} = \{q\} \cdot \{p\}$$

i.e. the multiplication of residue classes mod m is commutative.

Hence all the postulates for the ring are satisfied and so (R, +,.) forms a commutative ring a

Example 26:

Show that the set of all complex numbers of the form cos θ + i sin θ, where θ is any real number, forms a group with respect to the operation of multiplication of complex numbers.

Solution:

Closure Property: Let G = { z : z = cos θ + i sin θ, $\theta \in$ **R**}. **R** denotes the set of real numbers.

Associativity: For closure property, if z_1 = cosθ_1 + i sin θ_1 and z_2 = cos θ_2 + i sin θ_2 be any two members of G, then

$$z_1 z_2 = (\cos\theta_1 + i\sin\theta_1)(\cos\theta_2 + i\sin\theta_2)$$
$$= \cos(\theta_1 + \theta_2) + i\sin(\theta_1 + \theta_2)$$

which is also a member of G because $\theta_1, \theta_2 \in \mathbf{R} \Rightarrow \theta_1 + \theta_2 \in \mathbf{R}$.

Associativity Property: Multiplication of complex numbers is associative.

The complex number cos 0 + i sin 0 = 1 is a member of G and is the multiplicative identity.

Existence of Identity Inverse: If z = cos θ + i sin θ is any member of G, then

$$\cos(-\theta) + i\sin(-\theta) = \cos\theta - i\sin\theta$$

is also a member of G because $\theta \in \mathbf{R} \Rightarrow -\theta \in \mathbf{R}$.

Existence of Inverse: We have (cos θ + i sin θ) (cos θ + i sin θ) = $\cos^2\theta + i\sin^2\theta = 1$. So cos θ – i sin θ is the multiplicative inverse of cos θ + i sin θ.

Hence G is a group for multiplication of complex numbers.

Example 27:

Determine as to whether the following sets together with the indicated composition constitute groups:

1. *The set S of all ordered pairs (a, b) of real numbers for which $a \neq 0$ with respect to the operation $\times$ defined by*

   $(a, b) \times (c, d) = (ac, bc + d).$

2. *The set P (X) of all subsets of a non-empty set X, under the composition defined by the relation*

   $A * B = A \cup B \ \forall A \subseteq X, B \subseteq X.$

Solution:

Closure Property: Let (a, b) and (c, d) be any two members of X. Then $a \neq 0$ and $c \neq 0$. Therefore $ac \neq 0$. Consequently $(a, b) \times (c, d) = (ac, bc + d)$ is also a member of S. Hence S is closed with respect to the given composition.

Associativity: Let (a, b) , (c, d) and (e, f) be any three members of S. Then

$$[(a, b) \times (c, d)] \times (e, f) = (ac, bc + d) \times (e, f)$$
$$= ([ac]\, e, [be + d]\, e + f)$$
$$= (ace, bce + de + f).$$

Also $(a, b) \times [(c, d) \times (e, f)] = (a, b) \times (ce\ de + f)$

$$= (a\, [ce], b\, [ce] + de + f]$$
$$= (ace, bce + de + f).$$

Hence the given composition $\times$ is associative.

Existence of Left Identity: Suppose (x, y) is an element of S such that $(x, y) \times (a, b) = (a, b)\ \forall\ (a, b) \in S$.

Then $(xa, ya + b) = (a, g)$.

Hence $xa = a$ and $ya + b = b$.

These give $x = 1$ and $y = 0$. [Note that $a \neq 0$]

Therefore (1, 0) is the left identity.

Existence of let Inverse: Let (a, b) be any member os S. Let (x, y) be a member of S such that $(x, y) \times (a, b) = (1, 0)$.

Then $(xa, ya + b) = (1, 0)$. Hence $xa = 1$, $ya + b = 0$.

These give $x = 1/a$, $y = -\, b/a$.

Since $a \neq 0$, the therefore x and y are real numbers.

Also $x = \frac{1}{a} \neq 0$. Thus $\left(\frac{1}{a}, -\frac{b}{a}\right)$ is the left inverse of (a, b).

Hence S is a group.

Note: In the above group, we have

$$(a, b) \times (c, d) = (ac, bc + d)$$

and (c, d) × (a, b) = (ca, da + b).

Thus in general, (a, b) × (c, d) ≠ (c, d) × (a, b) *i.e.*, the composition is not commutative and hence the group is not abelian.

Closure Property: Let A and B be any two members of P (X). Then A ⊆ X, B ⊆ X. We have A * B = A ∪ B which is also a subset of X. Thus, A * B is also a member of P (X). Therefore P (X) is closed with respect to the given operation.

Associativity: We know that the union of sets is an associative operation.

Existence of left Identity: The empty set ∅ is a subset of X. Therefore ∅ is a member of P (X). If A is any member of P (X), we have ∅ * A = ∅ ∪ A. Therefore ∅ is the left identity.

Existence of Left Inverse: Let A be any non-empty member of P (X) *i.e.*, let A ⊆ X and A ≠ ∅. Now for every member S of P (X), we have S * A = S ∪ A ≠ ∅. Therefore no member S of P (X) can be the left inverse of A. Hence P (X) is not a group under the composition of union.

Example 28:

Show that the set V of all vectors (defined as directed line segments) forms an infinite abelian group with vector addition as composition.

Solution:

We know that the sum of two vectors is also a vector. Therefore V is closed with respect to addition of vectors. Also the vector addition is associative. If 0 is the zero vector, then 0 ∈ V and we have 0 + a = a = a + 0 ∀ a ∈ V. Therefore the vector 0 is the identity element.

If a ∈ V, then – a ∈ V and have – a + a = 0 = the identity element = a + (–a). Therefore – a is the inverse of a.

Also vector addition is commutative.

∴ V is an infinite abelian group with respect to addition of vectors.

Example 29:

Show that the set matrices

$$A_\alpha = \begin{bmatrix} \cos\alpha & -\sin\alpha \\ \sin\alpha & \cos\alpha \end{bmatrix}$$

where α is a real number, forms a group under matrix multiplication.

Solution:

Let G denote the set of matrices

$$A_\alpha = \begin{bmatrix} \cos\alpha & -\sin\alpha \\ \sin\alpha & \cos\alpha \end{bmatrix}, \text{ where } a \in R.$$

Hence R is the set of real numbers.

To prove that G is a group with respect to matrix multiplication.

Closure Property: Let A_α, B_β be any two elements of G while $\alpha, \beta, \in R$. We have

$$A_\alpha A_\beta = \begin{bmatrix} \cos\alpha & -\sin\alpha \\ \sin\alpha & \cos\alpha \end{bmatrix}\begin{bmatrix} \cos\beta & -\sin\beta \\ \sin\beta & \cos\beta \end{bmatrix}$$

$$= \begin{bmatrix} \cos(\alpha+\beta) & -\sin(\alpha+\beta) \\ \sin(\alpha+\beta) & \cos(\alpha+\beta) \end{bmatrix} = A_{\alpha+\beta} \in G, \text{ since } \alpha, \beta, \in R.$$

Associativity: Matrix multiplication is associative.

Existence of Left Identity: Since $0 \in R$. Therefore

$$A_0 = \begin{bmatrix} \cos 0 & -\sin 0 \\ \sin 0 & \cos 0 \end{bmatrix} \in G.$$ If A_α be any member of G, then

$$A_0 A_\alpha = A_{0+\alpha} \qquad [\because A_\alpha A_\beta = A_{\alpha+\beta}]$$

$$= A_\alpha.$$

$\therefore$ A_0 is the left identity.

Existence of Left Inverse: Let $A_\alpha \in G$. Then $A_{(-\alpha)} \in G$ because $\alpha \in R \Rightarrow -\alpha \in R$.

Now $A_{(-\alpha)} A_\alpha = A_{(-\alpha)+\alpha} \qquad [\because A_\alpha A_\beta = A_{\alpha+\beta}]$

$= A_0 =$ the left identity.

$\therefore$ $A_{(-\alpha)}$ is the left inverse of A_α.

Thus, each element of G possesses left inverse. Hence G is a group under matrix multiplication.

Example 30:

If G is a group and $a \in G$ is such that $aa = a$, prove that $a = e$.

Solution:

We have $aa = a$

$\Rightarrow$ aa = ae [$\because$ ae = a, e being the identity of G].

$\Rightarrow$ a = e, by left cancellation law in G.

Example 31(a):

Show that if every element of a group G is its own inverse, then G is abelian.

Solution:

Let a and b be any two elements of G. Then ab is also an element of G. Therefore $(ab)^{-1}$ = ab as it is given that every element of G is its own inverse.

Now $(ab)^{-1} = ab \Rightarrow b^{-1} = ab$

$\Rightarrow ba = ab$ [$\because a^{-1} = a, b^{-1} = b$].

Thus, we have ab = ba $\forall$ a, b, $\in$ G. Hence G is an abelian group.

Example 31(b):

Show that the set of all matrices $\begin{bmatrix} a & b \\ -b & a \end{bmatrix}$, *where a and b are real numbers not both equal to zero, is a group under matrix multiplication.*

Solution:

Let G denote the set of all matrices $\begin{bmatrix} a & b \\ -b & a \end{bmatrix}$, where a and b are real numbers not both equal to zero.

Closure Property: Let $A = \begin{bmatrix} a & b \\ -b & a \end{bmatrix}$, $B = \begin{bmatrix} c & b \\ -d & c \end{bmatrix}$ be any two members of G. Then a and b are real numbers not both equal to zero. Also c and d are real numbers not both equal to zero.

We have $AB = \begin{bmatrix} a & b \\ -b & a \end{bmatrix}\begin{bmatrix} c & b \\ -d & c \end{bmatrix} = \begin{bmatrix} ac - bd & ad + bc \\ -bc - ad & -bd + ac \end{bmatrix}$

$= \begin{bmatrix} p & q \\ -q & p \end{bmatrix}$, where p = ac – bd and q = ad + bc.

Obviously, p and q are real numbers and they cannot be both equal to zero as shown below,

We have det A = $a^2 + b^2 \neq 0$ and det B = $c^2 + d^2 \neq 0$.

But det (AB) = (det A). (det B).

Also det (AB) = $p^2 + q^2$

$\therefore$ det A $\neq 0$

and det B $\neq 0 \Rightarrow$ det (AB) $\neq 0$

$\Rightarrow p^2 + q^2 \neq 0$

$\Rightarrow$ at least one of p and q is not zero.

Thus p and q are real numbers not both equal to zero.

$\therefore$ A, b $\in$ G $\Rightarrow$ AB $\in$ G.

Hence the set G is closed for matrix multiplication.

Associativity: We know that matrix multiplication is associative.

Existence of Identity: The unit matrix $I = \begin{bmatrix} 1 & 0 \\ 0 & 1 \end{bmatrix}$ is obviously a member of the set G because it is obtained by putting a = 1 and b = 0. We have

$IA = A = AI, \forall A \in G.$

$\therefore$ the unit matrix I is the identity.

Existence of Inverse: Let $A = \begin{bmatrix} a & b \\ -b & a \end{bmatrix}$ be any member of the set G.

Then a and b are real numbers not both equal to zero.

We have det A = $a^2 + b^2 \neq 0$. Therefore the matrix A is non-singular and is therefore invertible. We must show that $A^{-1} \in G$. Let det A = m. Then m is a real number and m > 0.

We have $A^{-1} = \dfrac{1}{\det A} \text{adj } A = \dfrac{1}{m}\begin{bmatrix} a & -b \\ b & a \end{bmatrix} = \begin{bmatrix} a/m & -b/m \\ b/m & a/m \end{bmatrix}$ which is obviously a member of the set G because at least one of a/m and –b/m is not equal to zero.

Thus each member of G has its inverse in G.

Hence G is a group for matrix multiplication.

Example 31(c):

Show that the set C_0 of all non-zero complex numbers is a group with respect to multiplication of complex numbers.

Solution:

We have $C_0 = \{ a + ib : a \in \mathbf{R}, b \in \mathbf{R}$ and a and b are not both equal to $0\}$. The number 0 + i0 is the zero complex number.

Closure Property: We know that the product of two non-zero complex numbers is also a non-zero complex number. Therefore C_0 is closed with respect to multiplication.

Associativity: We know that multiplication of complex numbers is associative. Hence associative property.

Existence of Left Identity: We have $1 + i0 \in C_0$ since $1 + i0$ is a non zero complex number.

Also if $(a + ib) \in C_0$, then $(1 + i0)(a + ib) = a + ib$.

$\therefore$ $1 + i_0$ is the left identity. Hence $1 + 0i$ is the identifying element.

Existence of Left Inverse: Let $a + ib \in C_0$ *i.e.*, $a \in \mathbf{R}$, $b \in \mathbf{R}$ and a and b are not both equal to zero. Let $x + iy$ be the left inverse of $a + ib$. Then

$$(x + iy)(a + ib) = 1 + i0$$

$$\Rightarrow \quad (xa - yb) + i(xb + ya) = 1 + i0$$

$$\Rightarrow \quad xa - yb = 1,\ xb + ya = 0.$$

Solving the equations $xa + yb = 1$, $xb + ya = 0$, we get

$$x = \frac{a}{a^2 + b^2},\ y = \frac{-b}{a^2 + b^2}.$$

Since $a^2 + b^2 \neq 0$ and a and b are not both equal to zero, therefore x and y are real numbers not both equal to zero. Therefore

$$\frac{a}{a^2 + b^2} + i = \left(\frac{-b}{a^2 + b^2}\right) \in C_0$$

and it is the multiplicative left inverse of $a + ib$.

$\therefore$ C_0 is a group with respect to multiplication.

Since multiplication of complex numbers is commutative, therefore the group is an abelian group.

Example 31(d):

Prove that the set of all $m \times n$ matrices having their elements as integers (rational or real or complex numbers) is an infinite abelian group with respect to addition of matrices.

Solution:

Let M b the set of all $m \times n$ matrices with their elements as real numbers.

Closure Property: If $A \in M$, $B \in M$, then $A + B \in M$. The reason is that $A + B$ is also a matrix of the type $m \times n$ and the elements of the matrix $A + B$ are also real numbers since the sum of two real numbers is also a real number. Therefore M is closed with respect to addition of matrices.

Associativity: We know that addition of matrices is associative. Hence set M is associative w.r.t. addition.

Existence of Left Identity: If O be the null matrix of the type $m \times n$ then $O \in M$. Also if $A \in M$, we have $O + A = A$.

$\therefore$ The null matrix O is the left identity.

Existence of Left Inverse: If $A \in M$, the $-A \in M$ where $-A$ is the matrix whose elements are the negative of the corresponding elements of A. Also $-A + A = O =$ the left identity.

$\therefore$ $-A$ is the left inverse of A.

Also the addition of matrices is commutative and the set M is an infinite set

$\therefore$ (M, +) is an infinite abelian group

Example 32:

Let G be set of elements on which an algebraic operation $\times$ is defined such that $a \times b \in G$ for all a, b, $\in G$. Prove that G is an abelian group for this operation if the following postulates are satisfied:

1. $(a \times b) \times c = a \times (c \times b)$ *for all* $a, b, c \in G$;
2. *There exists a left identity* $e \in G$ *such that* $e \times a = a$ *for all* $a \in G$; *and*
3. *Corresponding to every element* $a \in G$, *there exists a left inverse* $a^{-1} \in G$ *such that* $a^{-1} \times a = e$.

Solution:

Let a, b be any two elements of G. By given postulate (2) left identity e is an element of G. Applying given postulate (1) for the elements e, a and b of G, we have

$$(e \times a) \times b = e\,(b \times a)$$

$$\Rightarrow \quad a \times b = b \times a$$

[$\because$ e is the left identity means $e \times a =$ and $e \times (b \times a) = b \times a$]

Thus, we have $a \times b = b \times a$ for all $a, b \in G$. Hence the operation $\times$ is commutative.

Now let $a, b, c \in G$. Then by (i), we have

$$(a \times b) \times c = a \times (c \times b)$$

$$= a \times (b \times c) \quad [\because b \times c = c \times b \text{ as just proved}]$$

Hence the operation $\times$ is associative. Thus G is an abelian group for the operation $\times$.

Example 33:

Show that the set G of all square matrices $[a_{ij}]_{n \times n}$ such that det $[a_{ij}]$ = ± 1 is a group under matrix multiplication. Show also that all those matrices in G for which det $[a_{ij}]$ = 1 form a group.

Solution:

Do yourself.

Hint: Here if A, B, ∈ G, then det A = ± 1 and det B = ± 1.

We have det (AB) = (det A). (det B) = ± 1 and so AB ∈ G.

If I is unit matrix of order n, then det I = 1 and so I ∈ G.

If A ∈ G, then det A = ± 1. Since det A ≠ 0, therefore the matrix A is non-singular and so is invertible. If B is the inverse of A, then

$$AB = 1 \quad \Rightarrow \det(AB) = \det I$$
$$\Rightarrow (\det A).(\det B) = 1$$
$$\Rightarrow \det B = \pm 1, \text{ Because } \det A = \pm 1.$$

Thus, A ∈ G ⇒ A^{-1} = B ∈ G. Thus each member of G has its inverse in G.

Hence G is a group for matrix multiplication.

Similarly we can show that all those matrices in G for which det $[a_{ij}]$ = 1 also form a group.

Example 34(a):

Show that the set

$$G = \{..., 2^{-4}, 2^{-3}, 2^{-2}, 2^{-1}, 1, 2, 2^2, 2^3, 2^4,...\}$$

forms an infinite abelian group with respect to multiplication.

Solution:

Closure Property: Let a, b be any two elements of G.

Then a = 2^r and b = 2^s

where r and s are some integers.

Now ab = $2^r 2^s = 2^{r+s}$ ∈ G since r + s is also some integer.

∴ G is closed with respect to multiplication. Hence closure property holding.

Associativity: We know that the elements of G are all rational numbers and the multiplication of rational numbers is associative.

Existence of Left Identity: We have $1 \in G$. If a is any element of G, then $1\ a = a$. Therefore 1 is the left identity. Hence every element has its inversing

Existence of Left Inverse: We have $a \in G \Rightarrow a = 2^r$ where r is some integer.

Now $2^r \in G \Rightarrow 2^r \in G$ since $-r$ is also some integer.

We have $2^{-r}\ 2^r = 2^0 = 1$. Therefore 2^{-r} is the left inverse of 2^r.

Commutativity: The multiplication of rational numbers is commutative.

Also the set G contains an infinite number of elements.

$\therefore$ G is infinite abelian group with respect to multiplication.

Example 34(b):

*Do the following sets form groups with respect to the binary operation * defined on them as follows:*

1. *the set Q_1 of all rational numbers other than 1, with the operation defined by $a * b = a + b - ab$.*
2. *the set Q' of all rational numbers other than -1, with the operation defined by $a * b = a + b + ab$.*

Solution:

Closure Property: Let $a, b \in Q_1$. Then a and b are rational numbers such that $a \neq 1$, $b \neq 1$.

Now $a * b = a + b - ab$ which is also a rational number and it cannot be equal to 1,

Since $a + b - ab = 1 \Rightarrow a + b - 1 = 0 \Rightarrow (a - 1)(1 - b) = 0$

$\Rightarrow a = 1$ or $b = 1$ which is not so.

$\therefore$ $a * b \in Q_1\ \forall\ a, b, \in Q_1$. Hence Q_1 is closed with respect to the given composition.

Associativity: If $a, b, c \in Q_1$, then

$$(a * b) * c = (a + b - ab) * c = (a + b - ab) + c - (a + b - ab)\ c$$
$$= a + b + c - ab - ac - bc + abc.$$

Also $a * (b * c) = a * (b + c - bc) = a\ (b + c - bc) - a\ (b + c - bc)$

$$= a + b + c - ab - ac + abc.$$

$\therefore$ $a + (b * c) = (a * b) * c\ \forall\ a, b, c \in Q_1$.

Existence of Left Identity: Let $e \in Q_1$ *i.e.*, let e be a rational number and $e \neq 1$. Then e will be the left identity iff $\forall\ a \in Q_1$ we have $e * a = a \Leftrightarrow e + a - ea = a \Leftrightarrow e - ea = 0 \Leftrightarrow e\ (1 - a) = 0 \Leftrightarrow e = 0$.

[Note that $a \in Q_1 \Leftrightarrow a \neq 1$]. Now $0 \text{ e} \in Q_1$; so 0 is the left identity.

Existence of Left Inverse: Let $a \in Q_1$ *i.e.* let a be a rational number and $a \neq 1$. Now $b \in Q_1$ will be the left inverse of a iff

$$b * a = 0 \Leftrightarrow b + a - ba = 0 \Leftrightarrow b\,(a - 1) = a$$

$$\Leftrightarrow b = \frac{a}{a-1}, \text{ since } a \neq 1.$$

Now $\frac{a}{a-1}$ is definitely a rational number. Also $\frac{a}{a-1}$ cannot be equal to 1. $\frac{a}{a-1} \in Q_1$ and so its is the left inverse of a.

Also $a * b = a + b - ab = b + a - ba = b * a$.

$\therefore$ The set Q_1 of all rational numbers except 1 is an infinite abelian group with respect to the given composition.

(2) proceed as in part (1). Here 0 is the identity element and the inverse of a is $-\frac{a}{a+1}$ which exists since $a + 1 \neq 0$ and $-\frac{a}{a+1} \neq -1$.

Example 35:

Show that the set $G = \{a + b\sqrt{2} : a, b \in Q\}$ is a group with respect to addition.

Solution:

Closure Property: Let x, y be any two elements of G. Then $x = a + b\sqrt{2}$, $y = c + d\sqrt{2}$ where $a, b, c, d, \in Q$.

Now $x + y = (a + b\sqrt{2}) + (c + d\sqrt{2}) = (a + c) + (b + d)\sqrt{2}$. Since $a + c$ and $b + d$ are elements of Q, therefore $(a + c) + (b + d)\sqrt{2} \in G$. Thus $x + y \in G \; \forall \; x, y \in G$. There fore G is closed with respect to addition.

Associativity. The elements of G are all real numbers and the addition of real numbers is associative.

Existence of Left Identity: We have $0 + 0\sqrt{2} \in G$ since $0 \in Q$. If $a + b\sqrt{2}$ is any element of G, then

$$(0 + 0\sqrt{2}) + (a + b\sqrt{2}) = (0 + a) + (0 + b)\sqrt{2} = a + b\sqrt{2}.$$

$\therefore$ $0 + 0\sqrt{2}$ is the left identity

Existence of Left Inverse: We have

$$a + b\sqrt{2} \in G \Rightarrow (-a) + (-b)\sqrt{2} \in G$$

since $\quad a, b \in Q \Rightarrow -a, -b \in Q$.

Now $\quad [(-a) + (-b)\sqrt{2}] + [a + b\sqrt{2}]$

$= [(-a) + a] + [(-b) + b] \sqrt{2} = 0 + 0 \sqrt{2} =$ then left identity.

$\therefore$ $(-a) + (-b) \sqrt{2}$ is the left inverse of $a + b \sqrt{2}$.

Hence G is a group with respect to addition.

Example 36:

Show that the set M of complex number z with the condition $|z| = 1$ forms a group with respect to the operation of multiplication of complex numbers.

Solution:

If $z = x + iy$ is a complex number, then

$$| z | = + \sqrt{(x^2 + y^2)} \text{ i.e., } | z |$$

is the non-negative value of the square root of $x^2 + y^2$.

Closure Property: Let $z_1, z_2 \in M$.

Then $| z_1 | = 1 \; | z_2 |$.

Since $| z_1 z_2 | = | z_1 | z_2 | = 1$, therefore $z_1 z_2 \in M$. Therefore M is closed with respect to multiplication of complex numbers.

Associativity: We know that multiplication of complex numbers is associative.

Existence of Left Identity: Since $| 1 + i0 | = 1$, therefore $1 + i0$ is an element of M. If $a + ib \in M$, we have $(1 + i0)(a + b) = a + ib$. Therefore $1 + i0$ is the left identity.

Existence of Left Inverse: Let $z = a + ib \in M$. Then $| z | = 1$. Since z is a non-zero complex number, therefore its multiplicative inverse $\frac{1}{z}$ exists. Let $u = \frac{1}{z}$. Then $| u | = \left|\frac{1}{z}\right| = \frac{1}{|z|} = \frac{1}{1} = 1.$ Therefore $u \in M$ and we have $uz = \frac{1}{z} . z = 1$. Thus u is the left inverse of z.

$\therefore$ M is a group with respect to multiplication of complex numbers.

Example 37(a):

*Does the set I of all integers form a group with respect to the binary operation * defined on it as follows:*

$$a * b = a + b + 1, \; \forall a, b, \in I.$$

Solution:

Closure Property: We have $a \in I, b \in \Rightarrow a + b + 1$ *i.e.* $a * b \in I$. Therefore I is closed with respect to the operation *.

Associativity: If a, b, c $\in$ I, then

$(a * b) * c = (a + b + 1) * c = (a + b + 1) + c + 1 = a + b + c + 2.$

Also $a * (b * c) = a * (b + c + 1) = a + (b + c + 1) + 1 = a + b + c + 2.$

$\therefore$ $(a * b) * c = a * (b * c)\ \forall$ a, b, c, $\in$ I. Hence associative property holding given composition

Existence of Left Identity: e $\in$ I will be the left identity if $e * a = a\ \forall$ a $\in$ I.

Now $\quad e * a = e + a + 1.$

$\therefore \quad e + a + 1 = a \Rightarrow e = -1.$

Since $\quad -1 \in I$ and we have for any $a \in I$

$(-1) * a = -1 + a + 1 = a,$

therefore -1 is the left identity element.

Existence of Left Inverse: If a $\in$ I, then b $\in$ I will be the left inverse of a if $b * a = -1$ (the left identity).

Now $b * a = -1 \Rightarrow b + a + 1 = -1$

$\Rightarrow b = -2 - a.$

Now $a \in I \Rightarrow -2 - a \in I.$

Also $(-2 - a) * a = (-2 - a) + a + 1 = -1.$

$\therefore$ $-2 - a$ is the left inverse of a.

Also $a * b = a + b + 1 = b + a + 1 = b * a$. Therefore the composition is also commutative.

Hence I is an infinite abelian group for the given composition.

Example 37(b):

Show that the set of all n $\times$ n non-singular matrices having their elements as rational (real or complex) numbers is an infinite non-abelian group with respect to matrix multiplication.

Solution:

A matrix A is said to be non-singular if $|A|$, *i.e.*, the determinant of the matrix A is not equal to zero.

Let M be the set of all n $\times$ n non-singular matrices with their elements as rational numbers.

Closure Property: Here we have $A \in M \Rightarrow A$ is of the type n $\times$ n, the elements of A are all rational numbers and $|A| \neq 0$. Similarly let $B \in M$. Now AB will be matrix of the type n $\times$ n, the elements of AB will all be rational number. Also | AB | = | A | | B |. Since | A | ≠ 0, and | B | ≠ 0, therefore | AB| ≠ 0. Thus AB ∈ M.

Associativity: We know that multiplication of matrices is associative.

Existence of Left Inverse: If that the unit matrix of the also n × n, then the elements of **I** are all rational numbers. Also | **I** | = 1 *i.e.*, ≠ 0. Therefore I ∈ M. If A ∈ M, we have IA = A. Therefore I is the left identity.

Existence of Left Inverse: We know the every non-singular matrix is inversible. Therefore if A∈M, there exists a non-singular matrix A^{-1} $\frac{1}{|A|}$ (Adj. A) with elements as rational numbers such that A^{-1} A = I = the left identity.

We know that multiplication of matrices is not in general commutative.

∴ M is an infinite non-abelian group with respect to multiplication of matrices.

Note: If M is the set of all n × n non-singular matrices with their elements as integers, then M is not a group with respect to matrix multiplication. The reason is that all such matrices are not necessarily inversible. For example the non-singular matrix $\begin{bmatrix} 1 & 2 \\ 3 & 2 \end{bmatrix}$ is not inversible.

We have $\begin{bmatrix} -\frac{1}{2} & \frac{1}{2} \\ \frac{3}{4} & -\frac{1}{4} \end{bmatrix}\begin{bmatrix} 1 & 2 \\ 3 & 2 \end{bmatrix} = \begin{bmatrix} 1 & 0 \\ 0 & 1 \end{bmatrix}$ = the identity element.

Since the elements of the matrix $\begin{bmatrix} -\frac{1}{2} & \frac{1}{2} \\ \frac{3}{4} & -\frac{1}{4} \end{bmatrix}$ are not integers, therefore this matrix does not belong to the set M.

Example 38:

If R is a ring with unity element 1, then prove that

(α) (–1) a = –a = a . (–1), ∀ a ∈ R;

(β) (–1) . (–1) = 1.

Solution:

We know if 1 be the unity element in the ring R, then

$$a \,.\, 1 = 1 \,.\, a = a \;\forall\; a \in R \qquad ...(i)$$

(α) Now $\quad [1 + (-1)] \,.\, a = 1 \,.\, a + (-1) \,.\, a, \;\forall\; a \in R$

or $0 . a = a + (-1) . a,$ from (i)

or $0 = a + (-1) -a$

or $(-1) . a = -a$

Again from (i) we get $a = a . 1$

or $a + a . (-1) = a . 1 + a.(-1)$, adding $a . (-1)$ on the right of both sides

or $a + a . (-1) = a . [1 + (-1)] = a . [0] = 0$

or $a . (-1) = -a$

(β) From part (α) we have $a. (-1) = -a$

Replacing a by (–1) in this result we get $(-1) . (-1) = - (-1)$

i.e. $(-1) . (-1)$ = additive inverse of (–1), where –1 is itself the additive inverse of 1 in R = 1.

Example 39(a):

If a, b belong to a ring R and $(a + b)^2 = a^2 + 2ab + b^2$, then show that R is a commutative ring.

Solution:

We have

$$(a + b)^2 = (a + b) . (a + b) = a . (a + b) + b . (a + b)$$

$$= a . a + a . b + b . a + b . b$$

or $(a + b)^2 = a^2 + ab + ba + b^2$ ...(i)

Now we are given that $(a + b)^2 = a^2 + 2ab + b^2$ ...(ii)

∴ from (i) and (ii) we conclude that ab = ba.

i.e. the ring R is a commutative ring.

Example 39(b):

Prove that the set of cube roots of unity is an abelian finite group with respect to multiplication.

Solution:

The set of cube roots of unity is, $G = \{(1, \omega\omega^2)\}$. Let us form the composition table as given below:

•	1	ω	ω^2
1	1	ω	ω^2
ω	ω	ω^2	$\omega^3 = 1$
ω^2	ω^2	$\omega^3 = 1$	$\omega^4 = \omega$

(G_1) **Closure axiom :** Since each element obtained in the table is a unique element of the given set G, multiplication is a binary operation. Thus the closure axiom is satisfied.

(G_2) **Associative axiom :** The elements of G are all complex numbers and we known that multiplication of complex numbers is always associative. Hence associative axiom is also satisfied

(G_3) **Identity axiom :** Since row 1 of the table is identical with the top border row of elements of the set, 1 (the element of the extreme left of this row) is the identity element in G.

(G_4) **Inverse axiom :** The inverse of 1 ω, ω^2 are 1, ω^2 and ω respectively.

(G_5) **Commutative axiom :** Multiplication is commutative in G because the elements equi-distant with the main diagonal are equal each to each.

The number of elements in G is 3. Hence (G,.) is a finite group of order 3.

Example 39(c):

Show that multiplication is a binary operation on the set $A = \{1,-1\}$ but not on $B = \{1, 3\}$.

Solution:

We have

$1.1 = 1$, $1.(-1) = -1$, $(-1)(-1) = 1$ all $\in$ A.

Hence multiplication is a binary operation on A.

Since for $3 \in B$, $3.3 = 9 \notin B$, multiplication is not binary operation on B.

Example 39(d):

o is operation defined in Z such that $aob = a + b - ab$ for $a, b \in Z$. Is the operation o a binary operation in Z? If so, is it associative and commutative in Z?

Solution:

If $a, b \in Z$ we have $a + b \in Z$, $ab \in Z$

and $a + b - ab \in Z$.

$\therefore$ $aob = a + b - ab \in Z$.

$\therefore$ o is a binary operation in Z .

Since $aob = a + b - ab = b + a - ba = boa$, 'o' is commutative in Z.

Now, $(aob)oc = (aob) + c - (aob)c$

$= a + b - ab + c - ac - bc + abc$

and $ao(boc) = a + (boc) - a(boc)$

$= a + b + c - bc - ab - ac + abc$

$= a + b + c - bc - ab - ac + abc$

$\therefore$ (aob)oc = ao(boc) and hence o is associative in Z.

Example 40:

Given that S = {A, B, C, D} where A = f, B = {a}, C = {a, b}. D = {a, b, c,}, show that S is closed under the binary operations ∪ (union of sets) and ∩ (intersection of sets) on S.

Solution:

$A \cup B = \phi \cup \{a\} = \{a\} = B.$

Similarly, $A \cup C = C$, $A \cup D = D$ and $A \cup A = A$

Also, $B \cup B = B$, $B \cup C = \{a\} \cup \{a, b\} = \{a, b\} = C$,

$B \cup D = \{a\} \cup \{a, b, c\} = \{a, b, c\} = D$

$C \cup C = C$, $C \cup D = \{a, b,\} \cup \{a, b, c\} = \{a, b, c\} = D.$

Hence ∪ is a binary operation on S.

Again, $A \cap A = A$, $A \cap B = \phi \cap \{a\} = \phi = A$,

$A \cap C = A$, $A \cap D = A$

and $B \cap B = B$, $B \cap C = \{a\} \cap \{a, b\} = \{a\} = B$

$B \cap D = \{a\} \cap \{a, b, c\} = \{a\} = B$

$C \cap C = C$, $C \cap D = \{a, b\} \cap \{a, b, c\} = \{a, b\} = C.$

Hence ∩ is a binary operation on S.

Example 41:

Let S be a non-empty set and o be an operation on S defined by aob = a for a, b ∈ S. Determine whether o is commutative and associative in S.

Solution:

Since aob = a for a, b ∈ S and boa = b for a, b ∈ S.

aob ≠ boa.

$\therefore$ o is not commutative in S.

Since (aob)oc = aoc = a for a, b, c

ao(boc) = aob = a ∈ S

$\therefore$ o is associative in S.

Example 42:

Show that the operation o given by aob = a^b is a binary operation on the set of natural numbers N. Is this operation associative and commutative in N?

Solution:

N is the set of natural numbers and o is operation defined in N such that aob = a^b for a, b ∈ N. When a, b ∈ N, a^b = a × a × a ... b times is also a natural number and hence a^b ∈ N.

∴ o is binary operation in N.

Let a, b, c ∈ N, then

(aob)oc = $(aob)^c$ = $(a^b)^c$ = a^{bc} and ao(boc) = aob^c = a^b

∴ (aob)oc ≠ ao(boc) and o is not associative in N.

Since $a^b \neq b^a$, *i.e.*, aob ≠ boa, 'o' is not commutative in N.

Example 43:

o is operation defined in Z such that aob = a + b – ab for a, b ∈ Z. Is the operation o a binary operation in Z? If so, is it associative and commutative in Z?

Solution:

If a, b ∈ Z we have a + b ∈ Z, ab ∈ Z

and a + b – ab ∈ Z.

∴ aob = a + b – ab ∈ Z.

∴ o is a binary operation in Z .

Since aob = a + b – ab = b + a – ba = boa, 'o' is commutative in Z.

Now, (aob)oc = (aob) + c – (aob)c = a + b – ab + c – ac – bc + abc

and ao(boc) = a + (boc) – a(boc) = a + b + c – bc – ab – ac + abc

= a + b + c – bc – ab – ac + abc

∴ (aob)oc = ao(boc) and hence o is associative in Z.

Example 44:

Prove that the set {1.–1, i. –i} is an abelian multiplicative finite group of order 4.

Solution:

Let G = {1.–1, i. –i}. The following will be the composition table for (G,.)

•	1	–1	i	–i
1	1	–1	i	– i
–1	–1	1	– i	1
i	i	– i	–1	1
– i	– i	i	1	–1

(G_1) **Closure axiom :** Since all me entries in the composition table are elements of the set G, the set G is closed under the operation multiplication. Hence closure axiom is satisfied.

(G_2) **Associative Axiom :** Multiplication for complex numbers is always associative.

(G_3) **Identity axiom :** Row 1 of the table is identical with that at the top border, hence the element 1 in the extreme *left column heading* row 1 is the identity element.

(G_4) **Inverse axiom :** Inverse of 1 is 1. *Inverse* of –1 is –1. Inverse of i is – i and of – i is i. Hence, inverse axiom is satisfied in G.

(G_5) **Commutative axiom :** Since in the table the 1st row is identical with 1st column, 2nd row is identical with 2nd column, 3rd row is identical with the 3rd column and 4th row is identical with the 4th column, hence the multiplication in **G** is commutative.

The number of elements in **G** is 4. Hence **G** is an abelian finite group of order 4 with respect to multiplication.

Example 45:

*Prepare the composition table for the set **G** = {1. – 1} with respect to ordinary multiplication. Show that **G** is a finite abelian group of order 2 with respect to ordinary multiplication composition.*

Solution:

The composition table is given below :

•	1	1
1	1	–1
–1	–1	1

(G_1) **Closure Property :** The set G is closed under multiplication as all the elements of the table belong to the set G.

(G_2) **Associative axiom :** Since each element in G is an integer, the multiplication is associative in G.

(G_3) **Existence of identity :** Since row I of the table is identical with the top row, the element 1 to the extreme left of this row is an identity element. Hence existence of identity verified.

(G_4) **Existence of inverse :** The inverse of 1 is 1 and that of – 1 is – 1. Hence existence of inverse is verified.

The number of elements in G is 2. Hence (G,.) is a finite abelian group of order 2.

Example 46:

Show that the operation o given by aob = a^b is a binary operation on the set of natural numbers N. Is this operation associative and commutative in N?

Solution:

N is the set of natural numbers and o is operation defined in N such that aob = a^b for a, b $\in$ N. When a, b $\in$ N, $a^b = a \times a \times a$ ... b times is also a natural number and hence $a^b \in$ N.

$\therefore$ o is binary operation in N.

Let a, b, c $\in$ N, then

$(aob)oc = (aob)^c = (a^b)^c = a^{bc}$ and $ao(boc) = aob^c = a^b$

$\therefore$ (aob)oc $\neq$ ao(boc) and o is not associative in N.

Since $a^b \neq b^a$, i.e., aob $\neq$ boa, 'o' is not commutative in N.

EXERCISES

1. If A = {1, – 1} and B = {1, 2}, then show that multiplication is a binary operation on A but not on B.
2. Introduce a symbol for each element of the symmetric group (ii) of degree 3, and construct a multiplication table for this group Show that n! is the order of S_n.
3. Prove that a group of order n with n $\leq$ 4 is necessarily Abelian.
4. Let m be a positive integer, consider the set

   $$I_m = (0, 1, \ldots, m-1\},$$

   and define the "product" of any two numbers in it to be the n mainder left when their ordinary product is divided by m. Construct a multiplication table similar to the non-zero elements of I_b. Does this set with this operation form a group Compute a similar table for the non-zero elements of I_7. Dot this system form a group?
5. Show that the group of all one-to-one mappings of a non-empty set X onto itself is non-Abelian if X has more than two elements.
6. Let G be a finite non-empty set which is closed under an associate multiplication with property. Show that G is a group.

7. Construct a multiplication table for the group of symmetries of square.
8. In the ring of even integers, why is 2 neither regular nor singular?
9. Consider the ring of all subsets of a non-empty set U, with the operations defined in Example 3. What are the regular elements in this ring? What are the singular elements? What are the divisors of zero? Under what conditions is this ring a field?
10. In each of the following rings of functions defined on the closed unit interval [0, 1], describe the regular elements, the singular elements, and the divisors of zero:

    (a) all real functions;

    (b) all continuous real functions;

    (c) all bounded continuous real functions.

    What changes are necessary in these descriptions it [0, 1] is replaced by (0, 1)?
11. Show that the ring Zn, is a field $\Leftrightarrow$ m is a prime number. (Hint: in showing that I_m is a field it m is prime, show first that in this case 7m has no non-zero divisors of zero, so that the non-zero elements of I_m are closed under multiplication and the cancellation law $ax = ay \Rightarrow x = y$ holds for these elements.
12. In the following, a set is given and binary operation is defined on it. Find whether the binary operation is commutative or associative or both in

    (i) $Z : a \text{ o } b = a + b + ab \ \forall \ a, b \in Z$

    (ii) $Z : a \text{ o } b = a + b + 1 \ \forall \ a, b \in Z$

    (iii) $S = \{a + b\sqrt{2} \mid a, b \in R\}$, usual multiplication

    (iv) S is the set of even integers : $a \text{ o } b = 3\,ab, \ \forall \ a, b \in S$.
13. Consider the binary operation o defined on the set A = {a, b, c, d} by the following composition table :

o	a	b	c	d
a	a	c	b	d
b	d	a	b	c
c	c	d	a	a
d	d	b	a	c

Compute:

(a) c o d and d o c

(b) b o d and d o b

(c) a o (b o c) and (a o b) o c

(d) Is o commutative? Associative?

14. Let R be a ring with identity, and show that R is a division ring ⇔ the non-zero elements of R form a group with respect to multiplication.

15. Let R be a ring with identity, and show that any divisor of zero in R is singular.

16. Q is the set of rational numbers, o is a binary operation defined on Q such that for a, b ∈ Q, a o b = a + b – ab. Show that (Q, o) is a semi-group.

17. Show that for the set of integers the operation o, where a o b = a^2b ∀ a, b ∈ I, is left distributive over ordinary addition (+) but not right distributive.

4

TREES AND PROBABILITY

INTRODUCTION TO TREES

What distinguishes all trees from other types of graphs is the absence of certain paths called *cycles*. Now, we will study a special type of relation that is exceptionally useful in a variety of computer science applications and that is usually represented by its digraph. These relations are essential for the construction of data bases and language compilers. They are called trees. Trees are among the most important graphs. This is mainly due to the fact that many of the applications of graph theory, directly or indirectly, involves trees.

Definition 1: *Cycle. A cycle is a circuit whose edge list contains no duplicates.*

The simplest example of a cycle in an undirected graph is a pair of vertices with two edges connected them. Since trees are cycle-free, we can rule out all multigraphs from consideration as trees.

Trees can either be undirected or directed graphs. We will concentrate on the undirected variety for now.

Definition 2: *Tree. An undirected graph is a tree if it is connected and contains no cycles or self-loops.*

Example 1:

(a) Graphs a, b and c in trees, while graphs d, e, and f are not trees.

(b) A K_2 is a tree. However, if $n \geq 3$, a Kn is not a tree.

(c) In a loose sense, a botanical tree is a mathematical tree. There are no cycles in the branch structure of a botaical tree.

(d) The structure of some chemical compounds are modeled by a tree. For example, butane consists of four carbon molecules and ten hydrogen molecules, where an edge between two molecules represents a bond between them. A bond is a force that keeps two molecules together. The same set of molecules can be linked together in a different tree structure to give us the compound isobutane. There are some compounds whose graphs are not trees. One example is benzene.

One type of graph that is not a tree, but is closely related, is a forest.

Definition: *Forest. A forest is an undirected graph whose components are all trees.*

Example 2:

The left-hand side can be viewed as a forest of three trees.

We will now examine several conditions that are equivalent to the one that defines a tree. The following theorem will be used as a tool in proving that the conditions are equivalent.

SPANNING TREES

The topic of spanning trees is motivated by a graph-optimization problem.

Example 1:

A map of Atlantis there are four campuses in its university system. A new phone system is being installed an the objective is to allow for communication between any two campuses to achieve this objective, the university must buy direct lines between certain edge for each direct phone line. Total communication is equivalent to G being a connected graph. This is due to the fact that two campuses can communicate over any number of lines. To minimize costs, the university wants to buy a minimum number of lines.

The solutions to this problem are all trees. Any graph that satisfies the requirements of the university must be connected, and if a cycle does exist, any line in the cycle can be delated, reducing the cost. Each of the sixteen trees that can be drawn to connect the vertices North. South, East, and West selves the problem as it is stated. Note that in each case, three direct lines must be purchased. There are two considerations that can help reduce the number of solutions that would be considered.

Objective 1: Give that the cost of each line depends on certain factors, such as the distance between the campuses, select a tree whose cost is as low as possible.

Objective 2: Suppose that communication over multiple lines is noisier as the number of lines increases. Select a tree with the property that the maximum number of lines that any pair of campuses must use to communicate with is as small as possible.

Typically, these objectives are not compatible; that is, you cannot always simultaneously achieve these objectives in all cases. In the case of the Atlantis university system, the solution with respect to Objective 1 is indicated with solid lines.

There are four solutions to the problem with respect to Objective 2: any tree in which one campus is directly connected to the other three. One solution with respect to Objective 2 is indicated with dotted lines. After satisfying the conditions of Objective 2, it would seem reasonable to select the cheapest of the four trees.

ROOTED TREES

In the next two sections, we will discuss rooted trees. Our primary focuses will be on general rooted trees and on a special case, ordered binary trees.

Informal Definition and Terminology: What differentiates rooted trees from undirected trees is that a *rooted tree* contains a distinguished vertex, called the *root.* Vertex A has been designated the root of the tree. If we choose any other vertex in the tree, such as M, we know that there is a unique path from A to M. The vertices on this path, (A, D, K, M), are described in genealogical terms:

M is a *Child* of K (so is L)

K is M's *parent*

A, D and K are M's *ancestors.*

D, K and M are *descendants* of A.

These genealogical relationship are often easier to visualize if the tree is rewritten so that children are positioned below their parents.

BINARY TREES

An *ordered rooted tree is* a rooted tree whose subtrees are put into a definite order and are, themselves, ordered rooted trees. An empty tree and a single vertex with no descendants (no subtrees) are ordered rooted trees.

Example 1:

(a) If we intend to apply the addition and subtraction operations in X first, we would parenthesize the expression to a * (b – c)/(d + e). Its expression tree appears.

(b) The expression trees for $a^2 - b^2$ and for (a + b) * (a – b) appear.

The three traversals of an operation tree are all significant. A binary operation applied to a pair of numbers can be written in three ways. One is familiar *infix*, form such as a + b for the sum of a and b. Another form is *prefix*, in which the same sum is written +*ab*. The final form is *postfix*, in which the sum is written *ab*+. Algebraic expressions involving the four standard arithmetic operations (+, –, *, and/) in prefix and postfix form are defined as follows:

Prefix: (a) A variable or number is a prefix expression.

(b) Any operation followed by a pair of prefix expressions is a prefix expression.

Postfix: (a) A variable or number is a postfix expression.

(b) Any pair of postfix expressions followed by an operation is a postfix expression.

The connection between traversals of an expression tree and these forms is simple.

(a) The preorder traversal of an expression tree will result in the prefix form of the expression.

(b) The postorder traversal of an expression tree will result in the postfix form of the expression.

(c) The inorder traversal of an operation tree will not, in general, yield the proper infix form of the expression. If an expression requires parentheses in infix form, an inordertraversal of its expression tree has the effect of removing the parentheses.

Example 2(a):

The trees are identical rooted trees, but as ordered trees, they are different.

If a tree rooted at v has p subtrees, we would refer to them as the first, second, ...pth subtrees. If we restrict the number of subtrees of each vertex to be less than or equal to two, we have a binary (ordered) tree.

Example 2(b):

For the tree the orders in which the vertices are visited are:

A-B-C-E-C-F-G, for the preorder traversal,

D-B-E-A-F-C-G, for the in order traversal, and

D-E-B-G-G-C-A, for the preorder traversal.

Example 2(c):

Binary Tree Sort. Given a collection of integers (or other objects than can be ordered), one technique for sorting is a binary tree sort. If the integers are $a_1, a_2, ..., a_n$ $n \geq 1$, *we first execute the following algorithm that creates a binary tree:*

(1) Insert a_1 into the root of the tree.

(2) For k: = 2 to n (* insert a_k into the tree*)

(2.1) r: = a_1

(2.2) inserted: = false

(2.3) While not inserted do

If a_k < r then

if v has a left child

then r: = left child of r

else make ak the left child of r and

inserted : = true

else (*$a_k \geq r$*)

if r has a right child

then r: = right of r

else make ak the right child of r and

inserted : = true

If the integers to be sorted are 25, 17, 9, 20, 33, 13, and 30, then the tree that is created is the one. The in order traversal of this tree is 9, 13, 17, 20, 25, 30, 33, the integers in ascending order. In general, the in order traversal of the tree that is constructed in the algorithm above will produce a sorted list. The preorder and postorder traversals of the tree have no meaning.

Example 3:

*The preorder traversal of the tree in Figure 10.4.5 is +–*ab/cde, which is the prefix version of expression X. The postfix traversal is ab * cd/–e +. Note that since the original form of X needed no parentheses, the inorder traversal, a * b – c/d + e, is the correct infix version.*

SOME PROPERTIES OF TREES

It is essential to know that the given graph is a tree or not, for this we will study certain properties of the trees. Some of them are as follows:

Theorem 1:

A Graph is a tree if and only if it is minimally connected.

Proof:

The vertices of a tree are connected together with the minimum number of edges. A Connected graph is said to be minimally connected if removal of any one edge from it disconnects the graph. A Minimally connected graph cannot have a circuit, otherwise, we could remove one of the edges in the circuit and still leave the graph connected. Thus a minimally connected graph is a tree.

Conversely, if a connected graph is not minimally connected, there must exist an edge e_k in G such that G - e_k is connected. Therefore, e_k is in some circuit, which implies that G is not a tree. Hence the theorem.

From the above theorems we can redefine a tree as follows:

A Graph G with n vertices is called a tree, if

(i) G is connected and is circuitless, or

(ii) G is connected and has n – 1 edges, or

(iii) G is circuitless and has n – 1 edges, or

(iv) There is exactly one path between every pair of vertices in G, or

(v) G is a minimally connected graph.

Theorem 2(a):

There is one and only one path between every pair of vertices in a tree, T.

Proof:

Since T is a connected graph, there must exist at least one path between every pair of vertices in T. Now, let there be two distinct paths between two vertices a and b in T. Clearly, the union of these two paths will contain a circuit and T cannot be a tree.

Theorem 2(b):

A Tree with n vertices has n-1 edges.

Proof:

We will prove this theorem by the method of induction on n, the number of vertices of T.

When n = 1, then T has only one vertex. Since it has no cycles, T cannot have any edge, i.e., it has $e = n - 1 = 1 - 1 = 0$.

Now, suppose the theorem is true for $n = m \geq 2$, where m is some positive integer. We will show that the result is true for $n = m + 1$. Let T be a tree with m + 1 vertices and let u v be an edge of T. Then if we remove the edge u v from T we get the graph T - uv. Then the graph is disconnected since T - uv contains no (u,v) path.

Thus, T – uv is disconnected. The removal an edge from a graph can disconnect the graph into at most two components. Let T – uv has two components T_1 and T_2. Since there were no cycles in T initially, both components are connected and without cycles. Thus, T_1 and T_2 are trees and each has fewer than n vertices. Now, we can apply the induction hypothesis to T_1 and T_2, to have

$$e(T_1) = v(T_1) - 1$$
$$e(T_2) = v(T_2) - 1$$

But, T_1 and T_2 are constructed by removing a single edge from T, hence

$$e(T) = e(T_1) + e(T_2) + 1$$

and $$v(T) = v(T_1) + v(T_2)$$

It follows that

$$e(T) = v(T_1) - 1 + v(T_2) - 1 + 1$$
$$= v(T) - 1 = m + 1 - 1 = m$$

Thus T has m edges.

Hence by the principle of mathematical induction the theorem is proved.

Theorem 2(c):

For any positive integer n, if G is a connected graph with n vertices and n–1 edges, then G is a tree.

Let n be a positive integer and let G be a particular arbitrarily chosen graph that is connected and has n vertices and n–1 edges. We know that a tree is a connected graph without cycles and as we have proved in theorem 3 that a tree has n–1 edges. Now, we have to prove that if G has no cycles and n -1 edges, then G is connected. Let us decompose G into m components, C_1, C_2, – - -, Cm. Each component is connected and it has no cycles since G has no cycles. Hence, each Cm is a tree.

Now, $$e_1 = n_1 - 1 \text{ and } \sum_{i=1}^{m} e_i = \sum_{u=1}^{m} (n_i - 1) = n - m$$

or $$e = n - m$$

Then it follows that m = 1 or G has only one component. Hence, G is a tree.

Theorem 3:

If in a graph G there is one and only one path between every pair of Vertices, G is a tree.

Proof:

Since there is a path between every pair of vertices, hence G is connected. A circuit in a graph implies that there is at least one pair of vertices a, b such that there are two distinct paths between a and b. Since G has one and only one path between every pair of vertices, G can have no circuit. Hence, G is a tree.

Pendant Vertices in a Tree

Each of the trees shown in the figures has several pendant vertices. A pendant vertex is defined as a vertex of degree one. The reason is that in a tree of n vertices we have n–1 edges and hence 2 (n–1) degrees to be divided among n vertices. Since no vertex can be of zero degree, we must have at least two vertices of degree one in a tree.

Distance and Centres in a Tree

In a connected graph G, the *distance* d (v_i, v_j) between two of its vertices v_i and v_j is the length of the shortest path, i.e. the number of edges in the shortest path, between them.

In a tree shown in Fig. 4.1 some of the paths are (a, b), (a, c), and (c, b). Lengths of these paths are d(a, b) = 1, d(a, c) = 2, d(c, b) = 1 etc.

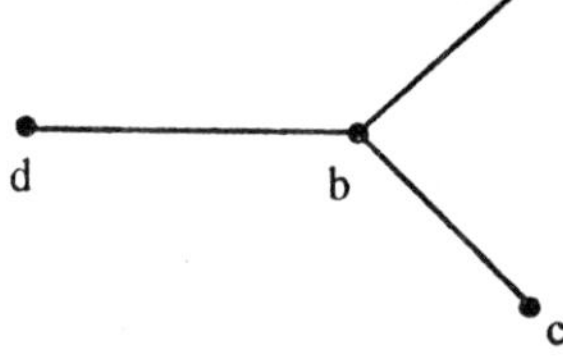

Fig. 4.1

A Metric. A function f (x, y), that satisfies following three conditions is called a *metric*.

(i) **Nonnegativity:** $f(x, y) \geq 0$, and $f(x, y) = 0$ if and only if $x = y$.

(ii) **Symmetry:** $f(x,y) = f(y, x)$

(iii) **Triangle inequality:** $f(x, y) \leq f(x, z) + f(z, y)$ for any z.

The distance $d(v_i, v_j)$ between two vertices of a connected graph evidently, satisfies conditions (i) and (ii). Since $d(v_i, v_j)$, is the length of the shortest path between vertices v_i and v_j, this path cannot be longer than another path between v_i and v_j which goes through a specified vertex v_k. Hence $d(v_i, v_j) \leq d(v_i, v_k) + d(v_k, v_j)$.

Therefore

The distance between vertices of a connected graph is a metric.

Eccentricity: The eccentricity E(v) of a vertex v in a graph G is the distance from v to the vertex farthest from v in G; i.e.,

$$E(v) = \max_{v_i \in G} d(v, v_i)$$

A vertex with minimum eccentricity in graph G is called a *centre* of G. The eccentricities of the four vertices in Fig. 4.1 are E(a) = 2, E(b) = 1, E(c) = 2 and E(d) = 2. Hence Vertex b is the centre of that tree.

TRAVERSING BINARY TREES

Traversing of a binary tree is a process to traverse the binary tree in a systematic way so that each vertex is visited exactly once. There are three standard ways of traversing a binary tree T with root. These three ways are called *preorder, inorder* and *post-order* and are as follows:

Preorder

(i) Process the root R.

(ii) Traverse the left subtree of R in preorder.

(iii) Traverse the right subtree of R in preorder.

In Order

(i) Traverse the left subtree of R in inorder.

(ii) Process the root R.

(iii) Traverse the right subtree of R in inorder.

Post Order

(i) Traverse the left subtree of R in post order.

(ii) Traverse the right subtree of R in post order.

(iii) Process the root R.

Observe that each of the three ways of traversing a binary tree contains the same three steps and the left subtree of R is always traversed before the rigt sub-tree. The difference between the methods is the time at which the root R is processed. In the pre order way, the root R is processed before the subtrees are traversed; in the In order, the root R is processed between the traversal of subtrees; and in the post order, the root R is processed after the subtrees are traversed.

These three ways are sometimes called the node-left-right (NLR) traversal, the left-node-right (LNR) traversal and the left-right-node (LRN) traversal respectively.

If only one order of traversal of a tree is given then it is possible to construct a number of binary trees. In this case a unique binary tree is not possible to be constructed. For construction of a unique binary tree we require two orders in which one has to be inorder, the other can be preorder or post order.

Traversal of a Unique Binary Tree When In Order and Post Order Traversal of the Tree is given

(i) The root of T is obtained by chosing the first vertex in its preorder.

(ii) First use the inorder Traversal to find the verticas in the left subtree of the binary tree. The left child of the root is obtained by selecting the first vertex in the preorder traversal of the left subtree. Draw the left child.

(iii) Use the inorder traversal to find the vertices in the right subtree of the binary tree. Then the right child is obtained by selecting the first vertex in the preorder traversal of the right subtree. Draw the right child.

(iv) The procedure is repeated until every vertex is not visited in preorder.

BINARY SEARCH TREES

A Binary search tree is basically a binary tree and is one of the most important data structures in computer science. Since it is a binary tree, hence it can be traversed in preorder, postorder, and inorder.

Definition : *A binary search tree is a binary tree T in which data are associated with the vertices. The data are arranged so that, for each vertex v in T, each data item in the left subtree of v is less than the data item in v and each data item in the right subtree of v is greater than the data item in v. Thus, a binary search tree for a set S is a labeled binary tree in which each vertex v is labelled by an element l(v) $\in$ S such that*

(i) for each vertex u in the left Subtree of v, $l(u) < l(v)$,

(ii) for each vertex u in the right subtree of v, $l(u) > l(v)$, and

(iii) for each element $a \in S$, there is exactly one vertex v such that $l(v)=a$.

The binary tree represented in Fig. 4.2 is a binary search tree.

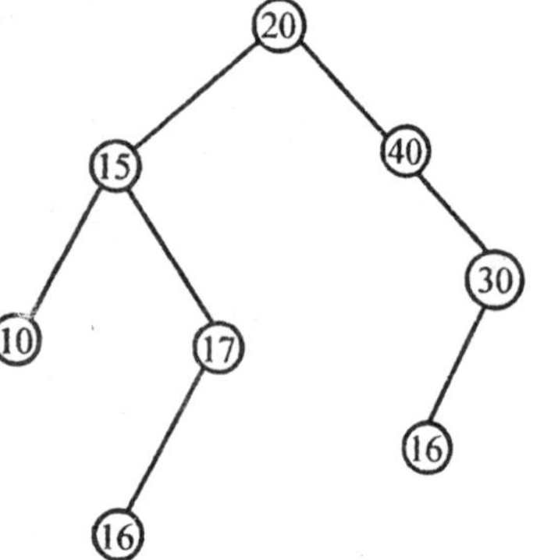

Fig. 4.2

Tracing a Binary Search Tree

The following procedure is used to form the binary search tree for a list of items.

(i) Create a vertex and place the first item in the list in this vertex and assign this as the key of the root.

(ii) To add a new item, first compare it with the keys of vertices already in the tree, starting at the root and moving to the left if the item is less than the key of the respective vertex if this has a left child, or moving to the right if the item is greater than the key of the respective vertex if this vertex has a right child.

(iii) When this item is less than the respective vertex and this vertex has no left child, then a new vertex with this item as its key is inserted as a new left child.

(iv) When the item is greater than the respective vertex and this vertex has no right child, then a new vertex with this item as its key then a new vertex has no right child, then a new vertex with this item as its key is inserted as a new right child.

In this way, we store all the items in the list in the tree and thus create a binary search tree.

ROOTED TREES

A Rooted tree is a tree in which a particular vertex is distinguished from the others and is called the *root*. In graph theory rooted trees are typically drawn with their roots at the top.

To form a rooted tree, first we place the root at the top. Under the root and on the same level, we place the vertices that can be reached from the root on the same path of length 1. Under each of these vertices and on the same level, we place vertices that can be reached from the root on a simple path of length 2. We continue in this way until the entire tree is drawn.

Level of a Vertex

The *level* of a vertex is the number of edges along the unique path between it and the root. The level of the root is defined as O.

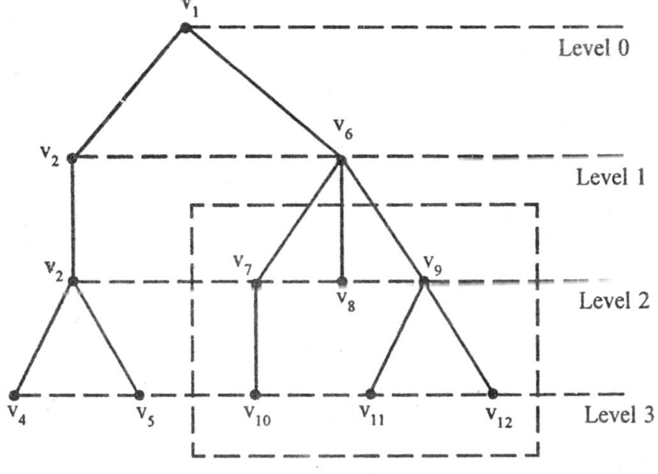

Fig. 4.3

The vertices immediately under the root are said to be in level 1 and so on. In Fig. 4.3, v_1 is the root, v_2 and v_6 are the vertices of level 1, v_3, v_7, v_8 and v_9 are the vertices of level 2 and so on.

Height

The *height* of the rooted tree is the maximum level to any vertex of the tree. The *depth* of a vertex v in a tree is the length of the path from the root v. For example in Fig. 4.3, height of the tree is 3 and depth of vertex v_3, is 2.

Children, Parent and Siblings

Given any internal vertex v of a rooted tree, the *children* of v are all those vertices that are adjacent to v and are one level farther away from than root than v. If w is a child of v, the v is called the *Parent* of w, and two vertices that are both children of the same parent are called *Siblings*. For example, in Fig. 4.3, v_4 is a child of v_3, v_3 is a parent of v_4, v_4 and v_5 are siblings.

If the vertex u has no children, then u is called a *leaf* or a *terminal vertex*. If u has either one or two children, then u is called an *internal vertex*.

Descendants and Ancestor

The *descendants* of the vertex v_i is the set consisting of all the children of v_i together with the descendants of those children. Given vertices v_j and v_k, if v_j lies on the unique path between v_k and the root, then v_j is an *ancestor* of v_k and v_k is a desendant of v_j.

m-ary Trees

A rooted tree is an *m-ary* if every internal vertex has at most m children. A m- ary tree is a *full m-ary* tree if every internal vertex has exactly m-children. In particular, the 2-ary tree is called *binary tree.* A *full binary tree* is a binary tree in which each internal vertex has exactly two children.

Theorem 1:

A full m- ary tree with i internal vertex has n = mi + 1 vestices.

Since the tree is a full m-ary, each internal vertex has m children and the number of internal vertex is i, the total number of vertex except in root is mi.

Therefore, the tree has n = mi + 1 vertices.

Let l be the number of leaves. Then n = l + i. Using the two equalities, we can deduce the following results:

A Full m- ary tree with

(i) n vertices has i = (n – 1)/m internal vertices and l = [(m – 1) n + 1]/ m leaves.

(ii) i internal vertices has $n = mi + 1$ vertices and $l = (m - 1) i + 1$ vertices and $l = (m - 1) i + 1$ leaves.

(iii) l leaves have $n = (ml - 1)/(m - 1)$ vertices and $i = (l - 1)/(m - 1)$ internal vertices.

Theorem 2:

There are atmost m^h leaves in an m- ary tree of height h.

For $h = 1$, the tree consists of a root with no more m children, each of which is a leaf. Hence there are $m^1 = m$ leaves in an m- ary of height 1.

We assume that the result is true for all m-ary trees of height less than h.

Let T be an m- ary tree of height h. The leaves of T are the leaves of subtrees of T obtained by deleting the edges from the root to each of the vertices of level 1. Each of these subtrees has height less than or equal to h -1. So by the method of induction, each of these rooted trees has at most m^{h-1} leaves. Since these are at most m such subtrees, each with a maximum of m^{h-1} leaves, there are at most $m. m^{h-1} = m^h$ leaves.

SPANNING TREES

A Tree T is said to be a *Spanning Tree* of a connected graph G if T is a subgraph of G and T contains all vertices of G. for example in Fig. 4.4, the subgraph in heavy lines is a spanning tree of the graph shown.

Since spanning trees are the largest trees among all trees in G, it is also called a *maximal tree subgraph* or *maximal tree* of G.

An edge in a spanning tree T is called a *branch* of T. An edge of G that is not in a given spanning tree T is called a *chord.* for example, edges a_1, a_2, a_3, a_4, a_5, and a_6, are branches of the spanning tree shown in Fig. 4.4, while edges b_1, b_2, b_3, b_4, b_5, b_6, and b_7 and b_8 are chords.

Branches and chords are defined only with respect to a given spanning tree. An edge that is a branch of one spanning tree T_1 may be a chord with respect to another spanning trace T_2.

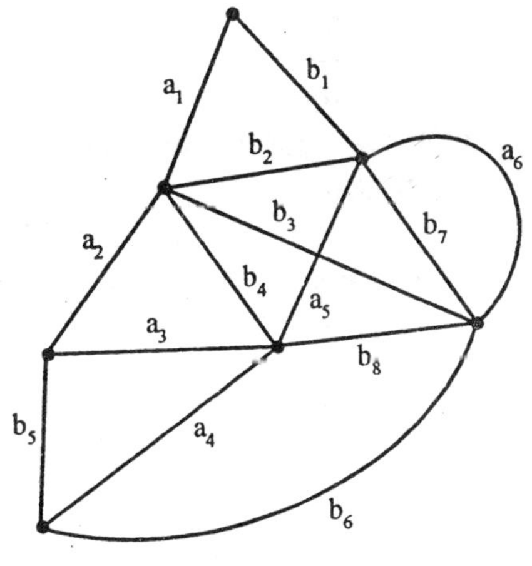

Fig. 4.4

Rank and Nullity

In a graph G, we define following numbers:

rank of G = number of branches in any spanning tree of G,

nullity of G = number of chords in G,

rank + nullity = number of edges in G.

The nullity of a graph is also referred to as its *cyclomatic number*, or *first Betti number*.

Minimal Spanning Tree

If graph G is a weighted graph, then the *weight of spanning tree* T of G is defined as the sum of the weights of all the branches in T. A spanning tree with the smallest weight in a weighted graph is called a *minimal spanning tree* or *shortest spanning tree* or *shortest - distance spanning tree.*

Algorithm for Shortest Spanning Tree

Kruskal's Algorithm

Kruskal's algorithm to find a shortest spanning tree in a given graph is as follows:

(i) List all edges of the graph G in order of nondecreasing weight.

(ii) Select a smallest edge of G.

(iii) For each successive step select another smallest edge that makes no circuit with the previously selected edges.

(iv) Continue until n – 1 edges have been selected, and these edges will constitute the desired shortest spanning tree.

Prim's Algorithm

(i) Draw n isolated vertices and label them v_1, v_2, ..., v_n

(ii) Tabulate the given weight of the edges of G in an n by n table. Set the weights of non-existent edges as very large.

(iii) Start from vertex v_1 and connect it to its nearest neighbour, i.e., to the vertex which has the smallest entry in row 1 of the table, say v_k.

(iv) Consider v_1 and v_k as are subgraph, and connect this subgraph to its closest neighbour, i.e., to a vertex other than v_1 and v_k that has the smallest entry among all entries in rows 1 and k. Let this new vertex is v_i.

(v) Regard the Tree with vertices v_1, v_k and v_i as one subgraph and continue the process until all n vertices have been connected by n 1 edges.

BINARY TREES

Binary Trees are a special class of rooted trees and are extensively used in the study of computer search methods, binary identification problems, and variable- length binary code.

A binary tree is a rooted tree in which each Vertex has atmost two children. Each child in a binary tree is designated either a *left child* or a *right child*, not both, and an internal vertex has at most one left and one right child. A *full binary* is a tree in which each internal vertex has exactly two children.

If an internal vertex v of a binary tree T be given, then the *left subtree* of v is the binary tree whose root is the left child of v, whose vertices consists of the left child of v and all its descendants, and whose edges consist of all those edges of T that connect the vertices of the left subtree together. The *right subtree* of v can be defined similarly.

Fig. 4.5(a) is a binary tree and Fig. 4.5(b) represents a full binary tree since each of its vertices has two children.

Theorem 1:

If T is a full binary tree with i internal vertices, then T has i + 1 terminal vertices and 2i + 1 total vertices.

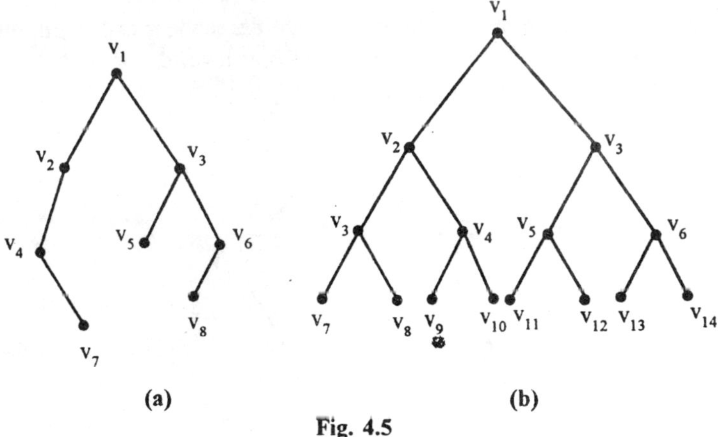

Fig. 4.5

The vertices of T consist of the Vertices that are children of some parent O, and the vertices that are not children of any parent. There is one nonchild—the root. Since there are i internal vertices, each parent having two children, there are 2i children, thus there are total 2i +1 vertices and the number of terminal vertices is (2i + 1) – i = i + 1.

Complete Binary Tree

If all the leaves of a full binary tree are at level d, then we call such a tree as a complete binary tree of depth d. A complete binary tree of depth 3 is shown in Fig. 4.6.

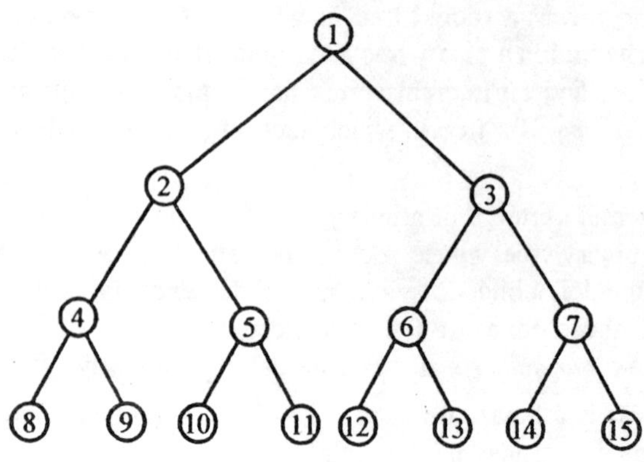

Fig. 4.6

Almost Complete Binary Tree

A binary tree of depth d is said to be almost complete binary tree if

(i) each node in the tree is either at level d or d-1,

(ii) For any node in the tree with a right descendant at level d, all the left descendants of this node are also at level d. Fig. 4.7 shows an almost complete binary tree.

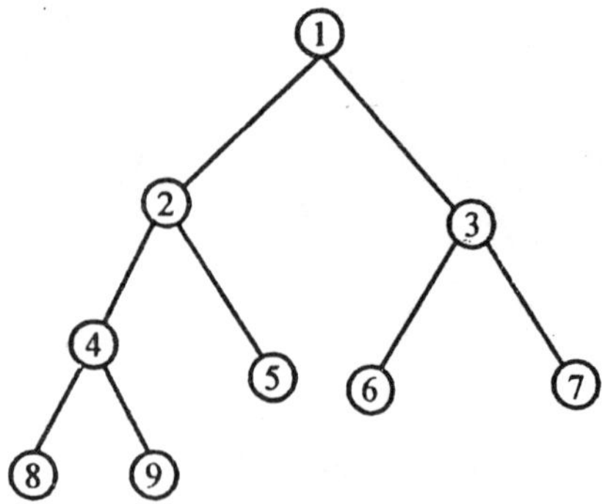

Fig. 4.7

PROBABILITY

The probability of a given event is an expression of likelihood or chance of occurrence of an event. A probability is a number which ranges from 0 (zero) to 1 (one). Zero for an event which can not occur and 1 for an event certain to occur. How the number is assigned would depend on the interpretation of the term 'Probability'. There is no general agreement about its interpretation and many people associate probability and chance with nebulous and mystic ideas. In this chapter we will make a systematic study of the theory of probability. But before discussing the theory of probability we shall explain some basic terms required in the study of probability. In every day life we talk about probability of occurrence of various events. Whenever there is an element of uncertainty about the occurrence of some events one makes a probability statement.

CENTRAL TENDENCY

After collecting the data for a random variable x and analyzing it in the form of frequency distribution, the next necessary step is to find the nature of distribution. In this context, a property for values of x to tend towards the centre is quite important. This property is called the *Central Tendency.* There are three most important measures of central tendency.

Mean : If x is the random variable, then its expected value E(x) itself is called the *mean* or *average* value of the random variable locates the middle of its probability function.

Mode : The mode of a random variable x is that value of the variable which occurs with the greatest frequency and is denoted by $\hat{x}$. It is possible that a particular distribution may not have a mode, or if it has a mode, it may not be unique.

A distribution is called unimodal, bimodal, trimodal, ... depending upon whether it has one, two, three ... modes.

For a discrete distribution, mode $\hat{x}$ is determined by the following inequalities:

$$P(x = x_i) \leq (x = \hat{x}),\ x_i \leq \hat{x}$$

and $$P(x = x_i) \leq (x = \hat{x}),\ x_j \leq \hat{x}$$

For a continuous distribution it is determined by the following equations:

$$\frac{d}{dx}\ [f(x)] = 0 \text{ and } \frac{d^2}{dx^2}\ [f(x)] < 0$$

Median : For a discrete or continuous distribution of a random variable x, the median is defined as the variate-value x satisfying the following in equations.

$$\left.\begin{aligned} P(x \leq X) &= \frac{1}{2} \\ \text{and} \quad P(x \geq X) &= \frac{1}{2} \end{aligned}\right\}$$

It is denoted by $\tilde{x}$

If a continuous distribution function has a probability density function f(x) in the range (a, b) then $\tilde{x}$ is given by

$$\int_a^{\hat{x}} f(x)\, dx = \frac{1}{2} = \int_b^{\tilde{x}} f(x)\, dx$$

Remark : If mean and median are known, then the mode can be calculated from the empirical formula

$$\text{Mean} - \text{Mode} = 3\ (\text{Mean} - \text{Median}).$$

THE SAMPLE SPACE

The set of all possible outcomes of a random experiment is called the sample space associated with the experiment.

The possible outcomes (elements of the sample space) are called *sample points.*

Example:

(i) Consider the random experiment of tossing a coin once. If we consider that the coin will never land on edge, there are two outcomes for the coin, viz., to land on :

(i) Head

(ii) Tail

Thus, if we denote by H, the occurrence of head, by T, the occurrence of tail and by S, the associated sample space, then

$$S = \{H, T\}.$$

(ii) If the random experiment consists of tossing a coin twice, the associated sample space is

$$S = \{HH, HT, TH, TT\}.$$

Discrete (Continuous) Sample Space

A sample space that consists of a finite (or an infinite but countable) number of sample points is called a discrete sample space.

A sample space which is not discrete is called a continuous sample space.

Example:

(i) Consider the number of customers arriving at a service counter through one hour. The sample space associated with this random experiment is:

$$S = \{0, 1, 2, ...\}$$

which is discrete

(ii) The sample space associated with the inter-failure time t of electronic component is given by $S = \{0 \le t < \infty\}$ which is continuous.

EVENT

An event is a subset of the sample space.

An event consisting of only a simple sample point is called elementary event.

Example:

Consider the random experiment of rolling a dice once. The associated sample space is

$$S = \{1, 2, 3, 4, 5, 6\}$$

Then, the subsets

$A = \{1, 2\}$, $B = \{2, 3, 4\}$ and $C = \{3\}$ etc. all represent events of S. Clearly. C is an elementary event,

Note :

(i) The sample space itself is an event.

(ii) We shall use capital letters to denote events and lower letters for sample points.

Equally Likely Outcomes

Outcomes of a random experiment are said to be equally likely if taking into consideration all the relevant evidences, there is no reason to expect one in preference to other.

Example:

(i) As the result of drawing a card from a well shuffled pack any card may appear in a draw. Thus the 52 possible outcomes are equally likely.

(ii) The falling and not falling down of a ceiling fan are not equally likely outcomes.

Favourable Outcomes : *The outcomes of a random experiment which entail the occurrence of an event A are said to be outcomes Favourable to A.*

Example:

In tossing a dice, the number of outcomes favourable to the event

$$E = \{\text{multiple of 2 appears}\}$$

is three, namely 2, 4, and 6.

Mutually Exclusive Outcomes : *The outcomes of a random experiment are said to be mutually exclusive if they cannot occur simultaneously.*

Example:

In the case of tossing a coin, H and T are mutually exclusive.

CLASSICAL DEFINITION OF PROBABILITY

If there are n exhaustive, pairwise mutually exclusive and equally likely outcomes of a random experiment and if m out of these are favourable to an event A, then the probability of happening of A, denoted by p(A), is defined by

$$p(A) = m/n \quad 0 \le m \le n, n > 0$$

Mathematical Definition of Probability : *If an event can happen in 'a' ways and fail in 'b' ways, and all these ways are equally likely to occur, then the probability of its happening is* $\frac{a}{a+b}$, *and that of its failing is* $\frac{b}{a+b}$.

Thus if p denotes the probability (or chance) of the happening of an event,

then $$p = \frac{a}{a+b}$$

and if q is the probability of its not happening, then

$$q = \frac{b}{a+b}$$

We have $$p + q = \frac{a}{a+b} + \frac{b}{a+b} = \frac{a+b}{a+b} = 1$$

i.e., $$p + q = 1.$$

Note : a + b the total number of ways and a are favourable ways.

$$\text{Hence } p = \frac{\text{favourable ways}}{\text{total number of ways}}$$

'Odds in Favour' and 'Odds Against'

Instead of saying that probability of happening of an event is $\frac{a}{a+b}$, we often say that odds are a to b is favour of the event or odds are b to a against the event.

ALGEBRA OF EVENTS

Equality of event : *Two event A and B are said to be equal if and only if they consist of exactly the same sample points.* We denote it by writing A = B.

Example:

Consider the random experiment of throwing a dice once. Then, the two events A and B defined by

A = An even number comes face up,

and B = {2, 4, 6}

are equal.

Union of two Events : *Let A and B be two events of some sample space. Then the set of all sample points that belong to either event A or event B (or both), is called the union of A and B and is denoted by $A \cup B$.*

In other words, *the occurrence of the event $A \cup B$ means that either event A occurs, or event B occurs (or both A and B occur), i.e., atleast one of A and B occurs.*

Example:

Consider the event,

$$A = \{1, 2\} \text{ and } B = \{2, 3, 4\} \text{ of}$$
$$S = \{1, 2, 3, 4, 5, 6\}$$

Then $A \cup B = \{1, 2, 3, 4\}$

Intersection of two Events : *Let A and B be two events of some sample space. Then, the set of all sample points that belong to both the events A and B is called the intersection of A and B* and is denoted by $A \cap B$.

In other words, *the occurrence of the event $A \cap B$ means that both the events, A and B occur simultaneously.* This is also called the *joint occurrence* of A and B.

Example:

In the preceding example,

$$A \cap B = \{2\}.$$

Impossible Event : *An event that contains no sample point is called an impossible event and is usually denoted by*

$$A = \phi \qquad \text{or} \qquad A = \{\ \}$$

In other words, *an impossible event can not occur.*

Example:

A ball is drawn at random from a bag containing 10 white and 5 black balls. Then the event

A = A red ball is drawn is an impossible event.

Complementary Event : *The event consisting of all sample points not contained in the event A, is called the complementary event of A* and is denoted by A'.

Example:

For any sample space

$$S' = \phi.$$

Mutually Exclusive Event : *Two events A and B are said to be mutually exclusive if $A \cap B = \phi$.*

In other words, two mutually exclusive event cannot occur simultaneously.

Example:

For the random experiment of tossing a coin, the two events defined by

A = Head appears; and B = Tail appears are mutually exclusive.

The events being the subsets of the sample space, satisfy certain laws of algebra of sets. These include

Commutative Laws

$$A \cup B = B \cup A; A \cap B = B \cap A$$

and associative laws,

$$(A \cup B) \cup C = A \cup (B \cup C)$$

$$(A \cap B) \cap C = A \cap (B \cap C)$$

THE ADDITION LAW OF PROBABILITY

If the events E_1, E_2, E_3, ..., E_n are mutually exclusive in pairs, then,

$$P(E_1 \cup E_2 \cup E_2 \ldots \cup E_n)$$
$$= P(E_1) + P(E_2) + P(E_3) + \ldots + P(E_n)$$

or

If p_1, p_2, ... p_n be separate probabilities of n mutually exclusive events, then the probability of the happening of any one of these events is given by

$$p = p_1 + p_2 + p_3 + ... + p_n.$$

Note : Probability of occurrence of at least one of the *two non-mutually exclusive events* is given by

$$P (A \cup B) = P(A) + P(B) - P(A \cap B).$$

THE CONDITIONAL PROBABILITY

Consider two events E_1 and E_2 in a sample space S. Here E_1 represents the event that has occurred m_1 number of times in n (large) number of trials, and E_2 represents the event that has occurred m_2 number of times out of these m_1 number of occurrence of E_1. Therefore, by using the classical definition of probability, we can find the probability of the combined happening of E_1 and E_2 in the same trial as

$$P (E_1 \cap E_2) = \frac{m_2}{n} = \frac{m_1}{n}, \frac{m_2}{m_1} \quad ...(i)$$

Obviously, $m_1/n = p(E_1)$, but the relative frequency m_2/m_1 can approximately be taken as the *conditional probability of occurrence* of event E_2 given that E_1 has occurred $[P(E_1) \neq 0]$; which is denoted by $P(E_2|E_1)$.

Now (i) becomes

$$P(E_1 \cap E_2) = P(E_1) P (E_2|E_1) \quad ...(ii)$$

i.e., the probability that both E_1 and E_2 occur is equal to the probability that E_1 occurs times the probability that E_2 occurs given that E_1 has occurred.

The above discussion suggests that we define the *conditional probability* of E_2 given E_1,

$$P(E_2|E_1) = P(E_1 \cap E_2)/P (E_1), P(E_1) > 0 \quad ...(iii)$$

Similarly, we can bet

$$P(E_1|E_2) = P(E_1 \cap E_2)/P (E_2), P(E_2) > 0 \quad ...(iv)$$

Conditional probability of E_2 statistics the following four properties :

(i) $o \leq P(E_2|E_1) \leq 1$.

(ii) If E_2 is an event which can not occur, then $P(E_2|E_1) = 0$

(iii) If the event E_2 is the entire sample space S, then $P(S|E_1) = 1$.

(iv) If E_2 and E_3 are two independent events in S, then

$$P(E_2 \cap E_3|E_1) = P(E_2|E_1) + P(E_3|E_1)$$

In case the occurrence of E_1 does not affect the occurrence of E_2, we have

$$P(E_2|E_1) = P(E_2) \quad ...(v)$$

Thus, E_1 and E_2 are independent if and only if,

$$P(E_2 \cap E_1) = P(E_1)\,P(E_2) \quad \text{[using results of (ii) and (v)]}$$

The Law of Total Probability

For any n events $E_1, E_2, ..., E_n$.

$$P(E_1 \cap E_2 \cap E_3 ..., \cap E_n)$$
$$= P(E_1) . P(E_2|E_1) . P(E_3|E_1E_2) ... P(E_n|E_1\ E_2 ... E_{n-1})$$

If the events $E_1, E_2, ..., E_n$ are independent, then

$$P(E_1 \cap E_2 \cap E_3 ..., \cap E_n)$$
$$= P(E_1) . P(E2) .... P(E_n).$$

DISCRETE AND CONTINUOUS VARIABLES

Suppose we collect data for the size of families in a certain town. It is obvious that the number of members in each family would in whole numbers. Thus, for example, there would be no family with 3.5 members. Such type of variable is called a *discrete variable.*

Now, suppose we are interested in measuring the heights of a large number of plants and if our unit of measurement is very fine, there would be no point along the scale of measurement at which we may not find the height of a plant.

A variable which takes all possible values between the limits, say a and b, is known as a continuous variable.

MATHEMATICAL EXPECTATION OF A RANDOM VARIABLE

For a *discrete random variable X*, the expected value is denoted by E(x) which is just the sum of the products of the possible values the random variable X takes on and their respective associated probabilities.

In other words, if the discrete random variable X takes n mutually exclusive values $x_1, x_2, ..., x_n$, and the others, with respective probabilities $p_1, p_2, p_3, ... p_n$, the expected value of x is given by

$$E(x) = p_1x_1 + p_2x_2 + p_3x_3 + ... p_nx_n$$
$$= \sum_{i=1}^{n} p_ix_i \quad ...(1)$$

Similarly, for the continuous random variable, the expected value can also be obtained by the formula

$$E(x) = \int_{-\infty}^{\infty} x f(x)\, dx \quad ...(2)$$

where f(x) is the probability density function.

Note : The mean of X is also called the *mathematical expectation* of X and is denoted by E(X).

RANDOM VARIABLE (OR CHANCE VARIABLE)

Let us consider some experiment whose sample space is S = $\{e_1, e_2, e_3, ..., e_n\}$

Definition : A random variable X is a rule which associates uniquely a real number with every elementary even $e_i \in S$, i = 1, 2, 3, ..., n, i.e, *a random variable is a real valued function which maps the sample space on the real line.*

In other words a random variable is a real valued function defined over the sample space of an experiment, i.e., a variable whose value is a number determined by the sample point (out come of the experiment) of a sample space is called a random variable.

For example, let X be a random variable which is the number of heads obtained in two independent tosses of a fair coin.

S = {HH, HT, TH, TT}

Then, X (HH) = 2, X (HT) = 1, X (TH) = 1, X (TT) = 0.

$\therefore$ X can take values 0, 1, 2.

A random variable is also known as *stochastic variable*

If the range of a random variable is a discrete random variable, it is called a *discrete random variable.* On the other hand the range of *continuous random variable* will be a set of continuous real numbers.

DISCRETE PROBABILITY DISTRIBUTIONS

The Binomial Distribution

Bernoulli trials : A series of independent trials which can result in one of the two mutually exclusive possibilities success or failure - such that the probability of success (or failure) in each trial is constant, then such repeated independent trials are called as Bernoulli trials.

If we perform a series of n Bernoullian trials such that for each trial, p is the probability of success and q is the probability of failure (p + q = 1), then probability of r successes in a series of n independent trials.

$$P(r) = n_{C_r}\ p^r\ q^{n-r}$$

where p + q = 1, and r = 0, 1, 2, 3, ..., n.

This is known as *Binomial distribution.* The constant n and p (or q) which appear in the binomial distribution are called the *parameters.*

The probabilities of 0 success, 1 success, 2 successes, ..., n successes are nothing but the first, the second, the third, ..., the (n + 1) th terms in the binomial expansion $(q + p)^n$. For this reason, the distribution is called binomial distribution.

Properties

(i) Mean of Binomial distribution = np

(ii) Variance = n p q

(iii) S.D. = $+\sqrt{npq}$

The Poisson Distribution

The discrete probability distribution obeying the probability law

$$P(X = x) = \frac{e^{-\lambda} \lambda^x}{x!}, \; x = 0, 1, 2, 3, ... \infty.$$

is called a Poisson distribution with parameter λ.

Properties

(i) The mean and variance of Poisson distribution coincide and their common value is λ.

(ii) If x and y are two independent Poisson random variables with mean λ and μ respectively, then xy is also a Poisson random variable with mean $\lambda + \mu$. The is called the *additivity* property of Poisson random variables.

BAYE'S THEOREM

If there be a set of mutually exclusive and exhaustive random events A_1, A_2, A_3, ..., A_n where $P(A_i) \neq 0$, then for any arbitrary random event E, we have.

$$P(E) = \sum_{i=1}^{n} P(A_i)\, P(E|A_i)$$

and $$P(A_i|E) = \frac{P(A_i E)}{\sum_{i=1}^{n} P(A_i)\, P(E|A_i)}$$

REPEATED TRIALS

We have learnt from the problems on probability that an event is not sure to happen in a single trial. We may make repeated trials to materialize the desired event. But as the trials are repeated the probability of different events arising from it cannot always be easily calculated. Now, we will give some methods of finding the probability of events in repeated trials.

Use of Binomial Expansion

If the probability of happening of an event in one trial is p and that of not happening is q such that $p + q = 1$, then the probability of r successes in n trials is $^{n}C_{r}\, p^{r}\, q^{n-r}$ *which is $(r + 1)$ th term in the expansion of $(q + p)^n$ with the help of Binomial Theorem.*

Use of Multinomial Expansion

If a die has f faces marked with 1,. 2, ..., f, the probability of throwing a total p with n dice is given by coefficient of x^p in the expansion of $(x^1 + x^2 + x^3 + ..., x^f)$ n divided by f^n.

In other words, if there are n cards with numbers a, b, c ... marked on them respectively, and p of them are drawn with replacement after each drawn the probability that the sum of numbers on the cards drawn is m is

= the coeff. of x^m in the

expansion of $\dfrac{(x^a + x^b + x^c + ...)^p}{n^p}$

PROBABILITY DISTRIBUTION OF A RANDOM VARIABLE

Any rule which assigns probabilities to each of the possible values of a random variable is called a probability distribution.

For discrete random variables, the most obvious and commonly used method of specifying the rule is to indicate the probability for each value separately. The function p(x), defined as

$$p(x) = P(X = x)$$

is called the probability distribution function.

For continuous random variables, the situation is some what complicated because the range of possible values is uncountably infinite. In this case, the distribution is defined by the *probability density function* f(x) for the prescribed range of random variable x.

The *probability density function,* f(x), is a function which, when integrated between a and b, gives the probability that the random variable will assume a value between a and b. That is,

$$P(a \leq X \leq b) = \int_a^b f(x)\, dx$$

CONTINUOUS PROBABILITY DISTRIBUTIONS

The Normal Distribution

A continuous random variable x with the range space $(-\infty, \infty)$ having the probability density function

$$f(x) = \frac{1}{\sigma\sqrt{2\pi}} \cdot e^{-1/2\left(\frac{x-\mu}{\sigma}\right)^2}$$

whose $-\infty \le x \le \infty$; $-\infty \le \mu \le \infty$, $\sigma > 0$,

is said to possess a Normal distribution with parameters μ and σ, denoted by N (μ, σ^2). Here μ is the mean and σ is the standard deviation of the normal distribution.

If x is a normal random variable with mean μ and standard deviation σ, then the random variable $Z = \dfrac{X-\mu}{\sigma}$ is called the standard normal variate which has the normal distribution with mean 0 and S.D. 1 and is given by

$$f(Z) = \frac{1}{\sqrt{2\pi}} e^{-1/2Z^2}, \quad -\infty \le Z \ge \infty$$

Normal Probability Curve

$$f(x) = \frac{1}{\sigma\sqrt{2\pi}} \cdot e^{-1/2\left(\frac{x-\mu}{\sigma}\right)^2}, \quad -\infty \le x \le \infty.$$

(i) The normal probability curve $y = f(x)$ is perfectly symmetrical about the line $x = \mu$ and is asymptotic to x-axis

(ii) Normal probability curves with some mean and different variances differ in maximum ordinate. More the variance, lesser is the max ordinate.

(iii) Mean, median and mode of normal distribution coincide at $x = \mu$.

(iv) The mean and variance are m and σ^2 respectively.

(v) The mean deviation about the mean of normal distribution is about 4/5 of its standard deviation.

(vi) The curve has two points of inflexion at $x = \mu \pm \sigma$. Both of these points are equidistant from the mean.

(vii) Total area under the curve and above x - axis from $-\infty$ to ∞ is unity.

Area of the normal curve between $\mu - \sigma$ and $\mu + \sigma$ is 68.27%

Area between $\mu - 2\sigma$ and $\mu + 2\sigma$ is 95.45%

Area between $\mu - 3\sigma$ and $\mu + 3\sigma$ is 99.73%

(viii) The most probable limits for a normal variate are

$\mu \pm 3\sigma$, i.e., $P[\mu - 3\sigma \leq X \leq \mu + 3\sigma]$

(ix) The sum and difference of independent normal variables also has a normal distribution.

MEAN, VARIANCE AD STANDARD DEVIATION (S.D.)

If the random variable X assumes the discrete values $x_1, x_2, x_3, \ldots x_r$ with corresponding probabilities $p_1, p_2, p_3, \ldots p_r$ then,

Mean (expected value) = $\overline{x}$

$$= \frac{\sum_{i=1}^{r} p_i x_i}{\sum_{i=1}^{r} p_i} = \sum_{i=1}^{r} p_i x_i \qquad [\because \Sigma p_i = 1] \qquad \ldots(1)$$

$$\text{Var}(x) = \sigma_x^2 = \sum_{i=1}^{r} (x_i - \overline{x})^2 p_i$$

$$= \sum_{i=1}^{r} x_i^2 p_i - \overline{x}^2 \text{ (when } \overline{x} \text{ is in fraction)} \qquad \ldots(2)$$

$$\text{S.D.} = \sigma_x = \sqrt{\text{var}} \qquad \ldots(3)$$

In the case of continuous random variable X with probability density function f(x), we have

$$\text{Mean} = \overline{x} = \int_{-\infty}^{\infty} x f(x)\, dx \qquad \ldots(4)$$

$$\text{Var}(x)\ \sigma_x^2 = \int_{-\infty}^{\infty} (x - \overline{x})^2 f(x)\, dx \qquad \ldots(5)$$

$$= \int_{-\infty}^{\infty} x^2 f(x)\, dx - (\overline{x})^2 \qquad \ldots(6)$$

$$\text{S.D.} = \sigma_x = \left(\sqrt{\text{Var}}\right) \qquad \ldots(7)$$

SOLVED EXAMPLES

Example 1:

Given the preorder and inorder traversal of a binary tree, draw the unique tree.

Preorder: 6 2 b 1 3 a 4 5 c

Inorder: b 2 3 1 6 a 5 4 c

Soultion:

Here 6 is the first vertex in preorder traversal, thus 6 is the root of the tree. Using inorder traversal, left subtree of 6 consists of the vertices 6, 2,

3 and 1. Then the left child of 6 is 2, since 2 is the first vertex in the preorder traversal in the left subtree. Similarly, right subtree of 6 consists of the vertices a, 5, 4, and c, than the right child of 6 is a since a is the vertex in the preorder traversal in the right subtree.

Repeating the above process with each node, we obtain the required tree as shown in Fig 4.8.

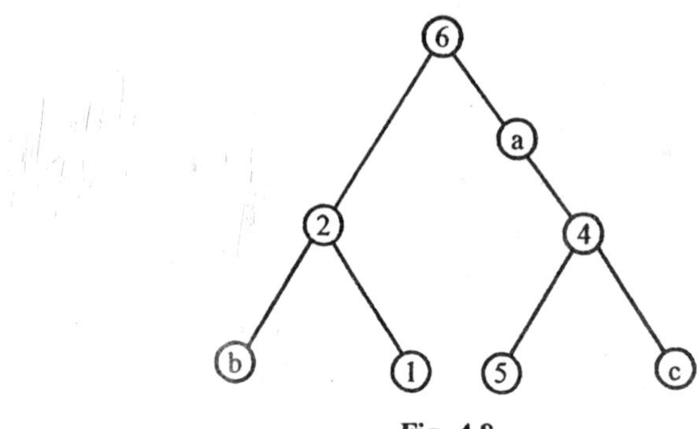

Fig. 4.8

Traversal of a Unique Binary Tree when Inorder and Postorder Traversal of the tree is given:

(i) The root of the binary tree obtained by choosing the last vertex in the postorder traversal.

(ii) To obtain the right child, we first use the inorder traversal to find the vertices in the rigth subtree. All the vertices right to the root vertex in the inorder traversal are the vertices of the right subtree. The right child of the root is obtained by selecting the last vertex in the post order traversal. Draw the right child.

(iii) Use the inorder traversal to find the vertices in the left subtree of the binary tree. Then the left child is obtained by selecting the last vertex in the post order traversal of the left subtree. Draw the left child.

(iv) The process is repeated until every vertex is not visited in the post order.

Example 2:

What is the chance of drawing 7 or 11 with two dice?

Solution:

7 can be obtained when, we get the following combinations:

(1, 6), (6, 1), (2, 5), (5, 2), (3, 4), (4, 3) i.e., in 6 different ways.

and 11 can be obtained when, we get

(5, 6), or (6, 5)

i.e, in 2 different ways.

Thus favourable ways of getting 7 or 11 = 6 + 2 = 8

And total number of ways of throwing 2 dice = $6^2 = 36$.

Hence the required chance $= \frac{8}{36} = \frac{2}{9}$

Example 3(a):

A bag contains 3 white and 2 black balls. Find probability of drawing a white ball.

Solution:

Here total number of ways in which a ball can be drawn = 5. Ways favourable for drawing a white ball = 3.

∴ Probability of drawing a white ball = 3/5.

Example 3(b):

The face cards (three from each suit) are removed from a full pack, out of the 40 remaining cards, 4 are drawn at random.

(a) What is the probability that they belong to different suits?

(b) What is the probability that 4 cards drawn belong to different suits and different denominations?

Solution:

Remaining 40 cards now contain 10 cards of each suit.

(a) Chance of drawing any card $= \frac{40}{40}$. After making the first draw, there remain 39 cards out of which 9 are of that suit a card of which has already been drawn.

∴ the chance of drawing a card of different suit of remaining 39 cards $= \frac{30}{39}$.

Similarly the chance of drawing a third card of another suit out of remaining 38 cards $= \frac{20}{38}$.

And chance of drawing fourth card of a different suit $= \frac{10}{37}$.

∴ The required chance

$$= \frac{40}{40} \times \frac{30}{39} \times \frac{20}{38} \times \frac{10}{37} = \frac{1000}{9139}$$

(b) Chance of drawing first card $= \frac{40}{40}$.

Now, there are 3 cards of the same value (denomination) and 9 cards of the same suit in the remaining 39 cards.

Therefore chance of drawing a card of different suit and different denomination $= \frac{27}{39}$.

Similarly chance of drawing a third card of different suit and different denomination $= \frac{16}{38}$.

And chance of a fourth card so drawn $= \frac{7}{37}$.

∴ the required chance $= \frac{40}{40} \times \frac{27}{39} \times \frac{16}{38} \times \frac{7}{37} = \frac{504}{9189}$

Example 3(c):

In shuffling a pack of cards, three are accidently dropped, find the chance that the missing cards should be from different suits.

Solution:

There are in all 52 cards. Chance of one dropping card being of any colour $= \frac{52}{52} = 1$

Since there are 39 cards of suits different to that a card which has already been drawn therefore the chance that the other card is of a different suit $= \frac{39}{51}$.

And lastly the chance that the third card is of different suit (since out of 50 remaining cards there are 26 cards of suits from which cards have not been drawn) $= \frac{26}{50}$

Hence the required chance

$$= 1 \times \frac{39}{51} \times \frac{26}{50} = \frac{169}{425}$$

Example 4:

A bag contains 3 red and 3 blue balls. Two draws of 2 balls each are made. Find the chance that the first draw gives 2 red balls and the second draw 2 blue balls.

(a) if the balls are returned to the bag after the first draw.

(b) if the balls are not returned.

Solution:

(a) *When the balls are returned.*

Total number of ways of drawing 2 balls from the bag = $^7C_2$, ways of drawing 2 red balls = $^4C_2$.

Therefore probability of drawing 2 red balls in the first draw

$$= \frac{^4C_2}{^7C_2} = \frac{4.3}{7.6} = \frac{2}{7} \quad ...(1)$$

But after the first draw the balls are replaced hence at the time of second draw the bag contains 4 red and 3 blue balls.

Therefore probability of drawing 2 blue balls in the second draw

$$= \frac{^3C_2}{^7C_2} = \frac{3.2}{7.6} = \frac{1}{7} \quad ...(2)$$

Both above drawings are independent. Hence the compound probability of drawing 2 red balls in the first and 2 blue balls in the second draw

$$= \frac{2}{7} \times \frac{1}{7} = \frac{2}{49}.$$

(b) *When the balls are not returned.*

As above probability of drawing 2 red balls in the first draw = 2/7.

But after this draw balls are not replaced, hence at the time of second draw there are 2 red and 3 blue balls in the bag.

Therefore the chance of drawing 2 blue balls in the second draw

$$= \frac{^3C_2}{^5C_2} = \frac{3.2}{5.4} = \frac{3}{10}$$

The above two drawings are dependent (compound).

Therefore the chance of drawing 2 red in the first draw and 2 blue in the second draw $= \frac{2}{7} \times \frac{3}{10} = \frac{3}{35}$.

Example 5:

A card is drawn from an ordinary pack and a gambler bets that it is a spade or an ace. What are the odds against his winning this bet?

Solution:

Total number of ways in which a card can be drawn from a pack of 52 cards = 52.

Ways in which a card can be spade or an ace = 13 + 3 = 16.

Hence the probability of gambler's winning the bet $= \frac{16}{52} = \frac{4}{13} = \frac{4}{9+4}$,

i.e., odds against his winning are 9 to 4.

Example 6:

Six cards are drawn at random from a deck of 52 cards. What is the probability that 3 will be red and 3 black?

Solution:

Total number of ways drawing 6 cards from a deck of 52 cards

$$= 52_{C_6}$$

Now, there are 26 red and 26 black cards. Therefore 3 red and 3 black cards. Therefore 3 red and 3 black cards can be drawn in $26_{C_3} \times 26_{C_3}$ ways.

Hence the required chance

$$= \frac{26_{C_3} \times 26_{C_3}}{52_{C_6}}$$

$$= \frac{\frac{26.25.24}{1.2.3} \times \frac{26.25.24}{1.2.3}}{\frac{52.51.50.49.48.47}{1.2.3.4.5.6}}$$

$$= \frac{13000}{39151} = 0.332$$

Example 7:

What is the chance that a leap year, selected at random will contain fifty-three Sundays?

Solution:

A leap-year has 366 days and therefore contains 52 complete weeks and 2 days over. Now these two days may make following seven combinations:

1. Monday and Tuesday
2. Tuesday and Wednesday,
3. Wednesday and Thursday,
4. Thursday and Friday,
5. Friday and Saturday,
6. Saturday and Sunday,
7. Sunday and Monday.

Of these total seven combinations, last two are favourable ways for having another Sunday.

$\therefore$ the required chance = 2/7.

Example 8(a):

A party of n persons sit at a round table, finds the odds against two specified individuals sitting next to each other.

Solution:

n persons can sit at a round table in $(n - 1)!$ different ways.

Let A and B be two specified persons. If A and B sit together, regarding AB as one item, the different arrangements in which AB occur together = $(n - 2)!$

But A, B can interchange their positions, hence favourable ways for A and B sitting next to each other

$$= 2\,(n - 2)!$$

$$\therefore \text{ the required chance } = \frac{2(n-2)!}{(n-1)!} = \frac{2}{n-1}$$

Therefore odds against are $n - 1 - 2$ to 2 or $n - 3$ to 2.

Example 8(b):

Eight letters, to each of which corresponds an envelope, are placed in the envelopes at random. What is the probability that all letters are not placed in the right envelopes.

Solution:

The total number of ways of placing 8 letters in 8 envelopes = 8!. Now all the letters can be placed in right envelopes only in 1 way.

$\therefore$ probability of placing the letters in right envelopes $= \frac{1}{8!}$ and probability that all letters are not placed in the right envelopes

$$= 1 - \frac{1}{8!}$$

Example 9:

A room has 3 lamps. From a collection of 10 light bulbs of which 6 are no good, a person selects 3 at random and puts them in the sockets. What is the probability that he will have light?

Solution:

There are 10 bulbs in all.

$\therefore$ Total number of ways of taking 3 out of 10

$$= 10_{C_3} = \frac{10 \times 9 \times 8}{1 \times 2 \times 3} = 120\,.$$

Now 6 bulbs are not good.

Hence ways of getting all the no good bulbs

$$= 6_{C_3} = \frac{6 \times 5 \times 4}{1 \times 2 \times 3} = 20$$

Hence the probability of getting all the no good bulbs, i.e., probability of not getting light $= \frac{20}{120} = \frac{1}{6}$.

Hence the probability of getting light $= 1 - \frac{1}{6} = \frac{5}{6}$.

Example 10:

The first twelve letters of the alphabet are written down at random. What is the probability that there are four letters between the A and the B?

Solution:

Let the twelve positions be denoted by

1, 2, 3, 4, 5, 6, 7, 8, 9, 10, 11, 12, when A is kept at 1, B must occupy place 6 to have four letters in between, when A is kept at 2, B must occupy place 7 etc. Lastly A may occupy a place 7 (so that B is at 12) just to include four letters in between.

Also A and B can interchange their position in 2! ways.

Thus A and B can take the described position in $2! \times 7 = 14$ ways.

Now, out of the remaining 10 alphabets four can be selected to put in between A and B in 10_{C_7} way; and these 4, and 6 others can give 4! 6! arrangements.

Thus the number of arrangements in which there are four letters between A and B = $14 \times 10_{C_3} \times 4!\ .\ 6!$.

And total number of arrangements in which twelve letters can be put = 12!

$\therefore$ The required probability

$$= \frac{14 \times 10_{C_4} \times 4! \times 6!}{12!}$$

$$= \frac{14 \times 10!}{12!} = \frac{14}{12 \times 11} = \frac{7}{66}$$

Example 11:

What is the chance in favour of throwing atleast 7 in a single throw with two dice?

Solution:

To throw atleast 7, the throw must give 7, 8, 9, 10, 11 or 12.

Now 7 can be made up with two dice in following ways:

(1, 6), (6, 1), (2, 5), (5, 2), (3, 4) and (4, 3) i.e., in 6 ways.

8 can be made up in following ways.

(2, 6), (6, 2), (3, 5), (5, 3), (4, 4) i.e., in 5 ways.

9 can be made up in following ways:

(3, 6), (6, 3), (4, 5), (5, 4), i.e., in 4 ways

10 can be made up when we get (4, 6)

(6, 4), (5, 5), i.e., in 3 ways.

11 can be made up when we get (5, 6)

(6, 5), i.e., in 2 ways.

12 can be made up when we get (6, 6), i.e., in 1 way only.

Hence that total number of ways of getting a number at least 7 = 6 + 5 + 4 + 3 + 2 + 1 = 2!

The total number of ways in which two dice can fall = $6^2 = 36$

Hence the required chance $= \frac{21}{36} = \frac{7}{12}$.

Example 12:

An urn contains n tickets bearing numbers from 1 to n and m tickets are drawn at a time. What is the probability that i of the tickets removed have numbers previously specified?

Solution:

The total number of ways in which m tickets can be drawn out of n tickets = ${}^{n}C_{m}$.

Out of n if i are specified tickets then n – i are unspecified tickets.

Ways of drawing i specified tickets from the previously specified tickets = ${}^{i}C_{i}$ and ways of drawing m – i tickets out of n – i other tickets = ${}^{n-i}C_{m-i}$

∴ The required probability

$$= \frac{{}^{i}C_{i} \times {}^{n-i}C_{m-i}}{{}^{n}C_{m}} = \frac{(n-i)!}{(m-i)!\,(n-m)!} \div \frac{n!}{m!\,(n-m)!}$$

$$= \frac{(n-i)!}{n!} \cdot \frac{m!}{(m-i)!}$$

$$= \frac{m\,(m-1)\ldots(m-i+1)}{n(n-1)\ldots(n-i+1)}$$

Example 13:

Given the post order and inorder traversal of a binary tree, draw the unique binary tree:

Post Order: p q n r m t u s l

In Order: p n q m r l t s u

Solution:

Here l is the last vertex in post order traversal, thus l is the root of the tree. Using inorder traversal, right subtree of root vertex l consists of the vertices t, s and u. The right child of l is s, since s is the last vertex in the post order traversal in the right subtree. SImilarly, left subtree of l consists of the vertices p, n, q, m and r.

Then the left child of l is m since m is the last vertex in the postorder traversal in the left subtree. Repeating the above process with each vertex, we obtain the required tree as shown in Fig. 4.9.

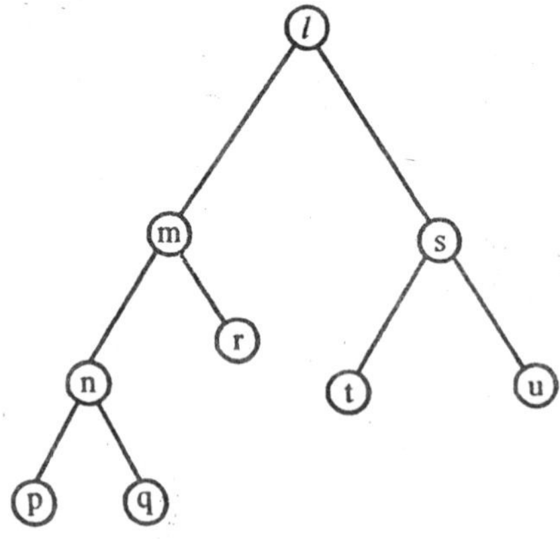

Fig. 4.9

Representation of Algebric Expressions by Binary Trees

Algebraic expressions can be represented by binary trees. Each variable or constant in the expression appears as an internal node in the tree whose left and right subtrees correspond to the operation at each vertex operates on its left and right subtrees from left to right.

Example 14:

Three groups of children contain respectively 3 girls and 1 boy; 2 girls and 2 boys; 1 girl and 3 boys. One child is selected at random from each group. Show that the chance that the three selected consist of 1 girl and 2 boys is 13/32.

Solution:

1 girl and 2 boys may be selected in following ways:

(i) girl from the 1st group, boys from 2nd and 3rd groups.

Probability of this event $= \frac{3}{4} \cdot \frac{2}{4} \cdot \frac{3}{4} = \frac{9}{32}$

(ii) girl from the 2nd group, boys from 1st and 3rd groups.

Probability of this event $= \frac{1}{4} \cdot \frac{2}{4} \cdot \frac{1}{4} = \frac{1}{32}$

(ii) girl from the 3rd group, boys from 1st and 2nd groups.

Probability of this event $= \frac{1}{4} \cdot \frac{2}{4} \cdot \frac{3}{4} = \frac{3}{32}$

All the above events are mutually exclusive hence the chance that any of these events happens $= \frac{9}{32} + \frac{1}{32} + \frac{3}{32} = \frac{13}{32}$.

Example 15(a):

A is one of 6 horses entered for a race, and is to be ridden by one of two jockeys B and C. It is 2 to 1 that B rides A in which case all the horses are equally likely to win. If C rides A his chance is trebled. What are the odds against his winning?

Solution:

The chance that jockey B rides the horse A = 2/3.

Now out of six all the horses are equally likely to win, hence the chance of A's success = 1/6.

Therefore the chances that A wins when B rides it

$$= \frac{2}{3} \cdot \frac{1}{6} = \frac{1}{9}.$$

Again the chance that C rides A = 1/3.

Now the chances A's winning being trebled $= 3 \times \frac{1}{6} = \frac{1}{2}$

The above two events are mutually exclusive

Hence A's chance of winning the race

$$= \frac{1}{9} + \frac{1}{6} = \frac{5}{18}$$

Odd against the success are 18-5 to 5 or 13 to 5.

Example 15(b):

An urn contains 10 white and 3 black balls. Another urn contains 3 white and 5 black balls. Two balls are transferred from the first urn and placed in the second and then one ball is taken from the latter. What is the probability that it is a white ball?

Solution:

First urn contains 10 white and 3 black balls, and second urn contains 3 white and 5 black balls.

Now there can be following possible events:

(i) The two balls taken from first urn are both white so that after placing them in the second urn, it contains 5 white and 5 black balls.

The probability of drawing a white ball from the second urn in this event

$$= \frac{10_{C_2}}{13_{C_2}} \cdot \frac{5_{C_1}}{10_{C_1}} = \frac{15}{52}$$

(ii) The two balls taken from the first urn are one white and one black, so that after placing them in the second urn, it contains 4 white and 6 black balls.

Probability of this event $= \frac{10_{C_1} \times 3_{C_1}}{13_{C_2}} \times \frac{4_{C_1}}{10_{C_1}} = \frac{2}{13}$

(iii) The two balls taken from the first urn are both black, so that after placing them in the second urn, it contains 3 white and 7 black balls.

Probability in this event $= \frac{3_{C_2}}{13_{C_2}} \cdot \frac{3_{C_1}}{10_{C_1}} = \frac{3}{26}$

All the three events are mutually exclusive.

Hence the required probability

$$= \frac{15}{52} + \frac{2}{13} + \frac{3}{26} = \frac{15 + 8 + 6}{52}$$

$$= \frac{29}{52}$$

Example 15(c):

A can hit a target 3 times in 5 shots, B, 2 times in 5 shots, C, 3 times in 4 shots. They fire a volley. What is the probability that 2 shots hit?

Solution:

2 shots can hit in the following events.

(i) A hits, B hits, C does not hit.

Probability of this event $= \frac{3}{5} \cdot \frac{2}{5} \cdot \frac{1}{4}$

$$= \frac{6}{100}$$

(ii) A hits, B does not hit, C hits.

Probability of this event $= \frac{3}{5} \cdot \frac{3}{5} \cdot \frac{3}{4} = \frac{27}{100}$

(iii) A does not hit, B hits and C hits.

Probability of this event $= \frac{2}{5} \cdot \frac{2}{5} \cdot \frac{2}{4} = \frac{12}{100}$

All the above three events are mutually exclusive.

Hence the probability that any one of them happens

$$= \frac{6}{100} + \frac{27}{100} + \frac{12}{100} = \frac{45}{100} = \frac{9}{20}$$

Example 16:

A husband and wife appear in an interview for two vacancies in the same post. The probability of husband's selection is 1/7 and that of wife's selection is 1/5. What is the probability that.

(i) only are of them will be selected?

(ii) both of them will be selected?

(iii) none of them will be selected?

Solution:

(i) P (only one of them will be selected)

$$= P(H)\,P(W') + P(W).\,P(H')$$

$$= \frac{1}{7} \times \frac{4}{5} + \frac{1}{5} \times \frac{6}{7} = \frac{10}{35} = \frac{2}{7}$$

(ii) P (both of them will be selected)

$$= P(H).\ P(W) = \frac{1}{7} \times \frac{1}{5} = \frac{1}{35}.$$

(iii) P (none of them will be selected)

$$= P(H').\ P(W') = \frac{4}{5} \times \frac{6}{7} = \frac{24}{35}.$$

Example 17:

The odds that a book will be favourably reviewed by three independent critics are 4 to 2, 4 to 3 and 2 to 3 respectively. What is the probability that of three reviews, a majority will be favourable.

Solution:

A majority of review will be favourable in the following events :

(i) 1st and 2nd critics favourably review, third goes against; the probability of this event $= \frac{3}{5} \cdot \frac{4}{7} \cdot \frac{3}{5} = \frac{36}{175}$

(ii) Ist and 3rd favourably review, 2nd goes against, the probability of this event

$$= \frac{3}{5} \cdot \frac{3}{7} \cdot \frac{2}{5} = \frac{18}{175}$$

(iii) 1st goes against, 2nd and 3rd favourably review; the probability of this event

$$= \frac{2}{5} \times \frac{4}{7} \times \frac{2}{5} = \frac{16}{175}$$

All the above three events are mutually exclusive

Hence the required probability

$$= \frac{36}{175} + \frac{18}{175} + \frac{16}{175} = \frac{70}{175}$$

$$= \frac{2}{5}$$

Example 18:

A's skill is to B's as 1 : 3; to C's as 3 : 2; and to D's as 4 : 3; find the chance that A in three trials, one with each person, will succeed twice at least.

Solution:

$$\text{A's chance of success over } B = \frac{1}{4},$$

$$\text{A's chance of success over } C = \frac{3}{5},$$

and $$\text{A's chance of success over } D = \frac{4}{7},$$

A will succeed twice at least in the following events :

(i) He wins over all the three, probability of this event

$$= \frac{1}{4} \times \frac{3}{5} \times \frac{4}{7} = \frac{3}{35}$$

(ii) He wins over B and C and not over D, probability of this event

$$= \frac{1}{4} \times \frac{3}{5} \times \frac{3}{7} = \frac{9}{140}$$

(iii) He wins over B and D and not over C, probability of this event

$$= \frac{1}{4} \times \frac{2}{5} \times \frac{4}{7} = \frac{2}{35}$$

(iv) He wins over C and D and not over B, probability of this event

$$= \frac{3}{4} \times \frac{3}{5} \times \frac{4}{7} = \frac{9}{35}$$

All the above four events are mutually exclusive.

Hence probability of succeeding twice at least (happening any one of the above four events.)

$$= \frac{3}{35} + \frac{9}{140} + \frac{2}{35} + \frac{9}{35} = \frac{12 + 9 + 8 + 36}{140}$$

$$= \frac{65}{140} = \frac{13}{28}.$$

Example 19:

Two cards numbered 1, two numbered 2, two numbered, 3 and two numbered 4 are placed in a box. After the box is shaken four men and their wives draw a card from the box. What is the probability that the cards drawn by each married couple bear the some numbers?

Solution:

There are in all 8 cards.

Let the pairs be denoted as

$$\begin{array}{cccc} A & B & C & D \\ H_1W_1 & H_2W_2 & H_3W_3 & H_4W_4 \end{array}$$

Now H_1 can draw any card and therefore his draw is a certainity, probability = 1.

His wife W_1 has the other card bearing the same number for a favourable draw out of the remaining 7 cards.

Therefore the probability that W_1 should draw a card bearing same number as of H_1 = 1/7.

Hence probability that couple A gets cards that bear same number = 1 × 1/7

Similarly for couple B, the probability = 1 × 1/5

For couple C, the probability = 1 × 1/3

and For couple D, the probability = 1 × 1/1

All the above events are dependent (compound),

Therefore the compound probability of all the above events

$$= \left(1 \times \frac{1}{7}\right) \times \left(1 \times \frac{1}{5}\right) \times \left(1 \times \frac{1}{3}\right) \times (1 \times 1) = \frac{1}{105}$$

Example 20:

A problem in discrete mathematics is given to three students A, B, C, whose chances of solving it one $\frac{1}{2}, \frac{1}{3}, \frac{1}{4}$ respectively what is the probability that the problem will be solved?

Solution:

Probability that A fails to solve the problem $= 1 - \frac{1}{2} = \frac{1}{2}$,

Probability that B fails to solve the problem $= 1 - \frac{1}{3} = \frac{2}{3}$,

and Probability that C fails to solve the problem $= 1 - \frac{1}{4} = \frac{3}{4}$

All the above three events are independent.

Hence the probability that A, B, C, all fail to solve the problem

$$= \frac{1}{2} \times \frac{2}{3} \times \frac{3}{4} = \frac{1}{4}.$$

Therefore the probability that the problem will be solved $= 1 - \frac{1}{4} = \frac{3}{4}$,

Example 21:

It is 8 : 5 against a person who is 40 years old living till he is 70 and 4 : 3 against a person now 50 living till he is 80. Find the probability that one at least of these persons will be alive 30 years hence.

Solution:

The chance that the first person will die after 30 years hence $= \frac{8}{13}$, and

the chance that the second person will die after 30 years hence $= \frac{4}{7}$.

Therefore the probability that both these die after 30 years hence $= \frac{8}{13} \times \frac{4}{7} = \frac{32}{91}$

and the probability that both of these will not die 30 years hence (i.e., at least one will be alive)

$$= 1 - \frac{32}{91} = \frac{59}{91}$$

Example 22(a):

A box contains 25 parts of which 10 are defective. Two parts are drawn at random from the box. What is the probability that both are good?

Solution:

Let E be the event the first part drawn is good', and F be the event 'the second part drawn is good'. Obviously, $E \cap F$ is the event 'both are good'. Therefore.

$$P(E \cap F) = P(E)\, P(F|E)$$

$$= \frac{15}{25} \cdot \frac{14}{24} = \frac{7}{20}.$$

Example 22(b):

A bag contains 5 white and 3 black balls and 4 are successively drawn and not replaced, what is the chance that they are alternately of different colours?

Solution:

Now balls of alternately different colours can be obtained in following two sets:

(i) White, black, white, black,

Probability of this event $= \frac{5}{8} \cdot \frac{3}{7} \cdot \frac{4}{6} \cdot \frac{2}{5} = \frac{1}{14}$

(ii) Black, white, black, white,

$$\text{Probability of this event} = \frac{3}{8}\cdot\frac{5}{7}\cdot\frac{2}{6}\cdot\frac{4}{5} = \frac{1}{14}$$

Both the above events are mutually exclusive.

$$\text{Hence the required chance} = \frac{1}{14} + \frac{1}{14} = \frac{1}{7}$$

Example 23:

'A' speaks truth in 60 percent cases and 'B' in 70% cases. In what percentage of cases are they likely to contradict each other in stating the same fact?

Solution:

It is clear that they will contradict each other only if one of them speaks the truth and other speaks a lie. The probability that A speaks the truth and B a lie

$$= \frac{60}{100} \times \frac{30}{100} = \frac{9}{50}.$$

The probability that B speak the truth and A a lie

$$= \frac{70}{100} \times \frac{40}{100} = \frac{14}{50}.$$

$$\therefore \text{ Required probability} = \frac{9}{50} + \frac{14}{50} = \frac{23}{50}.$$

∴ The % of cases in which they contradict each other

$$= \frac{23}{50} + 100 = 46\%$$

Example 24:

A lady buys a dozen eggs of which two turn out to be bad. She chooses four eggs to scramble for breakfast. Find the chance that she chooses:

(a) all good eggs

(b) three good and one bad,

(c) two good and two bad,

(d) atleast one bad egg.

Solution:

(a) The probability of having all good eggs

$$= \frac{10_{C_4}}{12_{C_4}} = \frac{210}{495} = \frac{14}{33}$$

(b) The probability of having 3 good and one bad

$$= \frac{10_{C_3} \times 2_{C_1}}{12_{C_4}} = \frac{120}{495} = \frac{16}{33}$$

(c) The probability of having 2 good and two bad eggs

$$= \frac{10_{C_2} \times 2_{C_2}}{12_{C_4}} = \frac{45}{495} = \frac{1}{11}$$

(d) The probability of having atleast one bad egg

$$= \frac{16}{33} + \frac{1}{11} = \frac{19}{33}$$

Example 25(a):

An experiment succeeds twice as often it fails. Find the chance that in the next six trials there will be at least four successes.

Solution:

Let p be the chance of success and q that of failure

$\therefore \quad p = 2q$ or $p = 2(1 - p)$ $\qquad [\because p + q = 1]$

or $\quad 3p = 2 \Rightarrow p = \frac{2}{3}$

$\therefore \quad q = 1 - \frac{2}{3} = \frac{1}{3}.$

Also, the probability of 6, 5, 4, ... successes are given by the successive terms in the expansion of $(q + p)^n$

i.e., $\left(\frac{1}{3} + \frac{2}{3}\right)^6$, since n = 6.

$\therefore$ Required probability of atleast four successes

$$= 6_{C_6}\left(\frac{2}{3}\right)^6\left(\frac{1}{3}\right)^{6-6} + 6_{C_5}\left(\frac{2}{3}\right)^5\left(\frac{1}{3}\right)^{6-5} + 6_{C_4}\left(\frac{2}{3}\right)^4\left(\frac{1}{3}\right)^{6-4}$$

$$= \left(\frac{2}{3}\right)^6 + 6\left(\frac{2}{3}\right)^5\left(\frac{1}{3}\right) + 15\left(\frac{2}{3}\right)^4\left(\frac{1}{3}\right)^2 =$$

$$= \frac{64 + 192 + 240}{3^6}$$

$$= \frac{496}{729}.$$

Example 25(b):

Four dice are thrown. What is the probability that sum of the numbers appearing on the dice is 18?

Solution:

From the multinomial theorem, the required probability

$$= \frac{1}{6^4} \text{ [Coeff. of } x^{18} \text{ in the expansion of } (x + x^2 + x^3 + x^4 + x^5 + x^6)^4]$$

$$= \frac{1}{6^4} \text{ [Coeff. of } x^{14} \text{ in the expansion of } (1 - x^6)^4 (1 - x)^{-4}]$$

$$= \frac{1}{6^4} \text{ [Coeff. of } x^{14} \text{ in the expansion of } (1 - 4x^6 + 6x^{12} + 4x^{18} + x^{21}] \times$$

$$(1 + 4x + 10x^2 + 20x^3 + 35x^4 + 46x^5 + 84x^6 + 120x^7 + 165x^8 + 220x^9 + 286x^{10} + 364x^{11} + 455x^{12} + 560x^{13} + 680x^{10} + ...)]$$

$$= \frac{1}{6^4} [680 - 4(165) + 6(10)]$$

$$= \frac{740 - 660}{6 \times 6 \times 6 \times 6} = \frac{5}{81}$$

Example 25(c):

There are three identical boxes containing respectively 1 white and 3 red balls, 2 white and 1 red ball, 4 white and 3 red balls. One box is chosen at random and two balls are drawn from it, if the balls white and red, what is the probability that they come from the second box?

Solution:

Let E be the event of drawing two balls one of which is white and the other red. Let the events of choosing 1st, 2nd and 3d boxes be A_1, A_2 A_3 respectively.

The $P(A_1)$ = probability of choosing 1st box out of three boxes

$$= \frac{1}{3} = P(A_2) = P(A_3).$$

Also $P(E|A_1)$ = p probability of drawing one white and the other red ball from the first box, when it has already been chosen.

$$= \frac{1}{4} \times \frac{3}{3} + \frac{3}{4} \times \frac{1}{3} = \frac{1}{2}.$$

Similarly $P(E|A_2) = \frac{2}{3} \times \frac{1}{2} + \frac{1}{3} \times \frac{2}{2} = \frac{2}{3}$

and $P(E|A_3) = \frac{4}{7} \times \frac{3}{6} + \frac{3}{7} \times \frac{4}{6} = \frac{4}{7}$

Now, the required probability

$$= P(A_2|E) = \frac{P(A_2) \,.\, P(E|A_2)}{P(E)}$$

$$= \frac{P(A_2) \,.\, P(E|A_2)}{P(A_2)\, P(E|A_1) + P(A_2)\, P(E|A_2) + P(A_3)\, P(E|A_3)}$$

$$= \frac{\frac{1}{3} \times \frac{2}{3}}{\frac{1}{3} \times \frac{1}{2} + \frac{1}{3} \times \frac{2}{3} + \frac{1}{3} \times \frac{4}{7}} = \frac{28}{73}$$

Example 26:

In a bolt *factory machines A, B, and C manufacture 25, 35 and 40 percent. Of the total of their output 5, 4 and 2 percent are defective. A bolt is drawn atrandom and is found to be defective. What is the probability that it was manufactured by machine B?*

Solution:

P(A) = probability that the bolt is manufactured by machine A

$$= \frac{25}{100} = \frac{1}{4},$$

Similarly $P(B) = \frac{35}{100} = \frac{2}{5},$

$P(C) = \frac{40}{100} = \frac{2}{5},$

Let D be the event of drawing a defective bolt,

Then the probability that a defective bolt being drawn from those manufactured by the machine A

$$= P(D|A) = \frac{5}{100} = \frac{1}{20}$$

Similarly $P(D|B) = \frac{4}{100} = \frac{1}{25}$ and

$P(D|C) = \frac{2}{100} = \frac{1}{50}.$

∴ The required probability

$$= P(B|D)$$

$$= \frac{P(B) . P(D|B)}{P(A).P(D|A) + P(B) . P(D|B) + P(C) . P(D|C)}$$

$$= \frac{\frac{7}{20} . \frac{1}{25}}{\frac{1}{4} . \frac{1}{20} + \frac{7}{20} . \frac{1}{25} + \frac{2}{5} . \frac{1}{50}} = \frac{28}{69}$$

Example 27:

In 2005 there will be three candidates for the position of principal– Dr Pant, Dr. Trivedi and Dr Singhal–Whose chances of getting the appointment are in the proposition 4 : 2 : 3 respectively. The probability that Dr. Pant if selected will make the uniform compulsory in the college is 0.3. The probability of Dr. Trivedi and Dr. Singhal doing the same are respectively 0.5 and 0.8. What is the probability that uniform will be compulsory is the college?

Solution:

Let the probabilities of Dr. Pant, Dr. Trivedi and Dr Singhal being appointed be A_1, A_2 and A_3 respectively and let the probability of uniform being compulsory be B. Then from Baye's Theorem, we have

$$P(B) = P(A_1)\ P(B|A_1) + P(A_2)\ P(B|A_2) + P(A_3)\ .\ P(B|A_3)$$

$$= \frac{4}{6} \times .3 + \frac{2}{9} \times .5 + \frac{3}{9} \times .8$$

$$= \frac{2}{15} + \frac{1}{9} + \frac{4}{15} = \frac{23}{45}$$

Example 28:

If ten fair coins are tossed, what is the probability that there are (a) exactly 3 heads (b) not more than 3 heads?

Solution:

The probability of head with a coin being 1/2. Probability that of 10 coins, r coins show heads is given by

$$P(r) = 10_{C_r} \left(\frac{1}{2}\right)^r \left(\frac{1}{2}\right)^{10-r}$$

$$r = 0, 1, 2, ..., 10$$

(a) $P(3) = 10_{C_r} \left(\frac{1}{2}\right)^3 \left(\frac{1}{2}\right)^{10-7}$

(b) Probability of not more than 3 heads

$= P(0) + P(1) + P(2) + P(3)$

$= \left(\frac{1}{2}\right)^{10} \left(10_{C_0} + 10_{C_1} + 10_{C_2} + 10_{C_3}\right)$

$= \frac{176}{1024} = \frac{11}{64}$

Example 29:

If the sum of the mean and the variance of a binomial distribution for 5 trials is 1.8, find the distribution.

Solution:

Mean = np and variance = npq

$\therefore$ np + npq = 1.8 $\Rightarrow$ np (1 + q) = 1.8

$\Rightarrow$ 5p [1 + (1 – p)] = 1.8 $\Rightarrow$ 5p (2 – p) = 1.8

$\Rightarrow$ $10p - 5p^2 = 1.8 \Rightarrow 5p^2 - 10p + 1.8 = 0$

$\Rightarrow$ $p^2 - 2p + 0.36 = 0$

$$\therefore\ p = \frac{2 \pm \sqrt{(-2)^2 - 4 \times 1 \times 0.36}}{2 \times 1}$$

$$= \frac{2 \pm \sqrt{(4 - 1.44)}}{2} = \frac{2 \pm 1.6}{2}$$

= 1 ± 0.8 - 1.8, 0.2.

But p cannot be equal to 1.8

$\therefore$ p = 0.2 = 1/5, q = 1 – 1/5 = 4/5

$\therefore$ n = 5, p = 1/5, and q = 4/5.

Example 30:

If the chance that any of the 5 telephone lines is busy at an instant is 0.01, what is the probability that all the lines are busy? What is the probability that not more than 3 lines are busy?

Solution:

$$p = 0.01 = \frac{1}{100},\ q = 1 - \frac{1}{100} = \frac{99}{100}$$

p (all lines are busy) = p(5)

$$= 5_{C_5}\left(\frac{1}{100}\right)^5\left(\frac{99}{100}\right)^{5-5} = \left(\frac{1}{100}\right)^5$$

p (not more than 3 lines are busy)

$$= 1 - [P(4) + P(5)]$$

$$= 1 - \left[5_{C_4}\left(\frac{1}{100}\right)^4\left(\frac{99}{100}\right)^{5-4} + 5_{C_5}\left(\frac{1}{100}\right)^5\left(\frac{99}{100}\right)^0\right]$$

$$= 1 - \left[5 \times \left(\frac{1}{100}\right)^4\left(\frac{99}{100}\right) + 1 \times \left(\frac{1}{100}\right)^5\right]$$

$$= 1 - \left[5 \times 99\left(\frac{1}{100}\right)^5 + \left(\frac{1}{100}\right)^5\right]$$

$$= 1 - (495 + 1)\left(\frac{1}{100}\right)^5 = 1 - (496)\left(\frac{1}{100}\right)^5$$

Example 31(a):

The number of customers arriving at a facility for service between 10 A.M. and 11 A.M. is a random variable, say X_1, with Poisson distribution with mean 2. Similarly, the number of customers arriving between 11 A.M. and 12 noon, say X_2, has a Poisson distribution with mean 6. If X_1 and X_2 are independent, find the probability that more than 5 customers will come between 10 A.M. and 12 noon.

Solution:

By additive property, $X = X_1 + X_2$ possesses a Poisson distribution with mean 8 (= 2 + 6).

Therefore the probability that there are x customers between 10 A.M. and 12 non is given by

$$P(X = x) = \frac{e^{-8}\, 8^x}{x!}, \; x = 0, 1, 2, 3, \ldots \infty.$$

Hence the probability that more than 5 customers arrive is

$$P\,(x > 5) = 1 - P\;(X \le 5)$$

$$= 1 - \sum_{x=0}^{5}\frac{e^{-8}\, 8^x}{x!}$$

$$= 1 - 0.191 = 0.809.$$

Example 31(b):

If a random variable has a Poisson distribution such that P(1) = P(2), find P(4). (use e^{-2} = 0.1353).

Solution:

$$p(r) = \frac{e^{-\lambda} \lambda^r}{r!}$$

$$\therefore \quad p(1) = e^{-\lambda}.\ \lambda, \qquad p(2) = \frac{e^{-\lambda}.\lambda^2}{2!}$$

But p(1) = p(2) (given)

$$\therefore \quad e^{-\lambda}.\lambda = \frac{e^{-\lambda}.\lambda^2}{2!}$$

$$\Rightarrow \quad \lambda = \frac{\lambda^2}{2} \Rightarrow \lambda = 2$$

$$\text{Now} \quad P(4) = \frac{e^{-2}.2^4}{4!} = \frac{0.1253 \times 16}{4 \times 3 \times 2 \times 1}$$

$$= \frac{2}{3}\ (0.1353) = 0.0902.$$

Example 32:

If the probability that an individual suffers a bad reaction from a certain injection is 0.001, determine the probability that out of 2000 individuals (a) exactly 3, (b) more than 2, individuals will suffer a bad reaction.

Solution:

Here $\quad p = 0.001, n = 2000$

$\therefore \quad \lambda = np = 2000 \times 0.001 = 2.$

Now, Poisson distribution is

$$P(r) = \frac{e^{-\lambda} \lambda^r}{r!}, r = 0, 1, 2, ...$$

$$\text{(a)} \quad P(3) = \frac{e^{-2}.2^3}{3!} = \frac{4}{3} e^{-2} = \frac{4}{3} \times 0.136 = 0.18$$

(b) P (more than 2) = P(3) + P(4) + P(5) + ...

$$= 1 - [P(0) + P(1) + P(2)]$$

$$= 1 - \left[\frac{e^{-2}.2^0}{0!} + \frac{e^{-2}.2^1}{1!} + \frac{e^{-2}.e^2}{2!}\right]$$

$= 1 - (1 + 2 + 2)\, e^{-2} = 1 - 5 \times 0.136$

$= 1 - 0.680 = 0.32.$

Example 33(a):

It is known from the past experience that the number of telephone calls made daily in a certain community between 3 p.m. and 4 p.m. have a mean of 352 and a S.D. of 31. What percentage of the time will these be more than 400 telephone calls made in this community between 3 p.m. and 4 p.m.?

Solution:

$$\mu = 352, \sigma = 31$$

$$Z = \frac{x - \mu}{\sigma} = \frac{400 - 352}{31} = \frac{48}{31} = 1.55$$

Now, $F(1.55) = 0.9394$ (From tables)

$\therefore$ Probability $= 1 - F(1.55) = 1 - 0.9394$

$= 0.0606$

$\therefore$ The required percentage of line

$= (0.0606 \times 100) = 6.06\%.$

Example 33(b):

The marks obtained by a large group of students in a final examination in discrete mathematics have a mean of 58 and a S.D. of 8.5. Assuming that these marks are approximately normally distributed, what percentage of students can be expected to have obtained marks from 60 to 69, both inclusive?

Solution:

Here $\mu = 58, \sigma = 8.5$

$$Z = \frac{X - \mu}{\sigma}$$

$$\therefore \quad Z_1 = \frac{60 - 58}{8.5} = \frac{2}{8.5} = 0.24$$

and $$Z_2 = \frac{69 - 58}{8.5} = \frac{11}{8.5} = 1.29$$

$F(Z_2) - F(Z_1) = F(1.29) - F(0.24)$

$= 0.9015 - 0.5948$ (From tables)

$= 0.3067$

∴ The percentage of students expected to have obtained marks from 60 to 69

$$= 0.3067 \times 100 = 30.67\%$$

Example 33(c):

During the course of a day a machine turns out either 0, 1, or 2 defective pens with probabilities 1/6, 2/3 and 1/6 respectively. Calculate the mean value and the variance of the defective pens produced by the machine in a day.

Solution:

The probability distribution in this example is given by :

x	:	0	1	2
p(x)	:	1/6	2/3	1/6

$$\therefore \quad \bar{x} = \sum_{i=0}^{2} p_i \, x_i = \frac{1}{6}.0 + \frac{2}{3}.1 + \frac{1}{6}.2 = 1$$

$$\text{Var}(x) = \sum_{i=0}^{2} x_i^2 \, p_i - \bar{x}^2$$

$$= \frac{1}{6}.0^2 + \frac{2}{3}.1^2 + \frac{1}{6}.2^2 - 1^2 = \frac{1}{3}$$

Example 34:

The probability density function for a continuous random variable x is given by

$$f(x) = \begin{cases} \frac{1}{2} \sin x\,dx & 0 \le x \le \pi \\ 0 & \text{otherwise} \end{cases}$$

Find the mean and variance.

Solution:

$$\text{mean} = \int_0^{\pi} x \left(\frac{1}{2} \sin x\right) dx$$

$$= \left[x\left(-\frac{1}{2}\cos x\right) - 1.\left(-\frac{1}{2}\sin x\right)\right]_0^{\pi}$$

or $\quad \bar{x} = \pi/2$

$$\text{Var}(x) = \int_0^{\pi} x^2 \left(\frac{1}{2} \sin x\right) dx - \bar{x}^2$$

Solution:

Here

$$f(x) = 1/2, 2 < x < 4$$
$$= 0, \text{ otherwise.}$$

The weight expected to be lost,

$$E(x) = \int_2^4 x.\frac{1}{2}\,dx = \left[\frac{x^2}{4}\right]_2^4$$
$$= 1/4\,[4^2 - 2^2] = 3 \text{ kg.}$$

EXERCISES

1. Construct a binary tree whose in-order and pre-order traversal is given below:

   In-order: 5, 1, 3, 11, 6, 8, 2, 4, 7

   Pre-order: 6, 1, 5, 11, 3, 4, 8, 7, 2

2 Construct a binary tree whose in-order and pre-order traversal is as follows:

   In-order: d, g, b, e, i, h, j, a, c, f

   Pre-order: a, b, d, g, e, h, i, j, c, f

3. Represent each of the following expressions in a binary tree:

   (a) (a + b) * (c - d)

   (b) (x + 7) * [(4 * y + z)/(s + 3)]

   (c) ((3 + x) - (4 * x)) - (x - 2)

   (d) (5 * a) + (3 -(6 * a))) + (a - (3 * b))

4. Draw the binary tree to represent the expression $(x + 3y)^5 (a - 2b)$.

5. If three cards are drawn from a pack of 52, what is the probability that all three will be kings?

6. From a pack of 52 cards two are drawn at random. Find the chance that one is a king and the other a queen.

7. A number is chosen from each of two sets

   1, 2, 3, 4, 5, 6, 7, 8, 9;   1, 2, 3, 4, 5, 6, 7, 8, 9;

   If p_1 denotes the probability that the sum of the two numbers be 10 and p_2 the probability that their sum be 8; find $p_1 + p_2$.

8. What is the chance that non-leap year should have fifty three Sundays?

9. There are four addressed envelopes. Find the chance that all the letters are not despatched in the right envelopes.
10. If four coins are tossed, find the chance that there should be two tails.
11. A and B stand in a ring with 10 other persons. If the arrangement of the 12 persons is at random, find the chance that there are exactly 3 persons between A and B.
12. A traffic census shows that out of 1000 vehicles passing a junction point on a highway 600 turned to the right. Find the probability of an automobile turning to the right at this junction.
13. If the chance of A winning a certain race be 1/6 and the chance of B winning it is 1/8. What is the chance that neither should win?
14. Given the tree with root at A as shown in Fig. 4.10

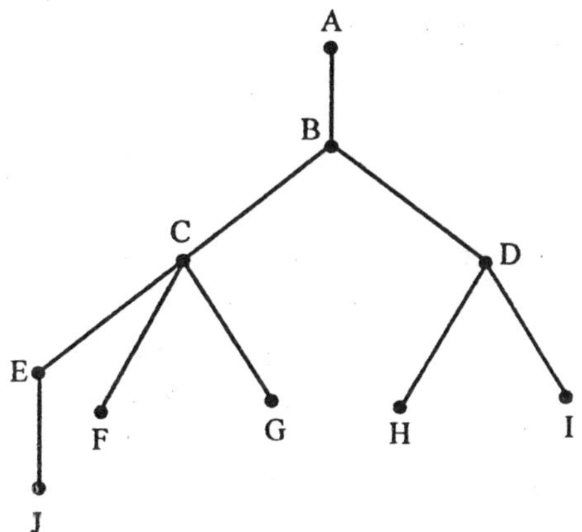

Fig. 4.10

(i) Find the children of D and E
(ii) Find the leaves,
(iii) Find the internal vertices,
(iv) Find the parents of C and H,
(v) Find the descendants of C and E
(vi) Find the siblings of F and H.

15. The chances of winning of two horses are 1/3 and 1/6 respectively. What is the probability that atleast one will win when the horses are running
(a) in different races
(b) the same race?

16. Draw all spanning trees of the graph shown below:

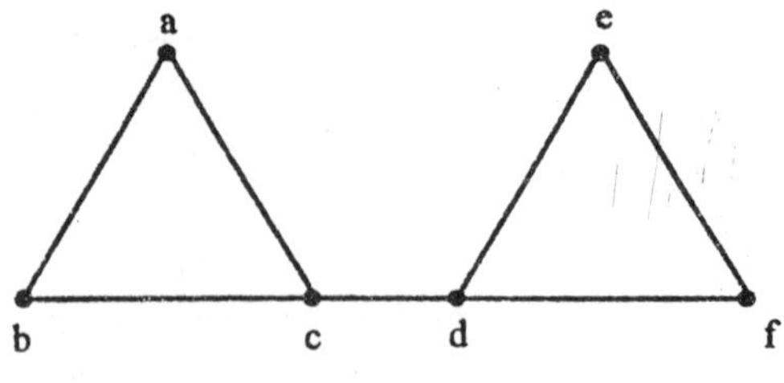

Fig. 11.21

17. A bag contains 6 white and 9 black balls. The drawing of 4 balls are made such that

    (a) the balls are replaced before the second draw

    (b) the balls are not replaced before the second draw.

    Find the probability that the first drawing will give 4 white and the second 4 black balls in each case.

18. An urn contains 10 white and 3 black balls. Another urn contains 3 white and 5 black balls. Two balls are transferred from the first urn and placed in the second and then one ball is taken from the latter. What is the probability that it is a white ball?

19. A can hit a target 4 times in 5 shots; B, 3 times in 4 shots, C twice in 3 shots. They fire a volley. What is the probability that two shots atleast hit?

20. There are two bags, one of which contains 5 red and 7 white balls and the other 3 red and 12 white balls. One ball is to be drawn from one or other of two bags. Find the chance of drawing a red ball.

21. The odds against a certain event are 5 to 2 and the odds in favour of another event, independent of the former are 6 to 5, find the chance that one at least of the events will happen.

22. The probability that a 50 years old man will be alive at 60 is .83 and the probability that a 45 years old woman will be alive at 55 is .87. What is the probability that a man who is 50 and wife who is 45 will be alive 10 years hence?

23. If an average one vessel in every 10 is wrecked, find the probability that out of 5 vessels, 4 atleast will arrive safely.

24. In tossing ten coins, what is the probability of getting exactly 5 heads?

25. An urn contains 5 balls. Two balls are drawn and are found to be white. What is the probability of all the balls being white?

26. One urn contains 3 white and 2 black balls, another contains 5 white and 3 black balls. If one of the urns is chosen atrandom and a ball taken out of it, find the probability that it is white.

27. Discuss and criticize the following

    P(A) = 2/3, P(B) = 1/4, P(C) = 1/6

    where A, B and C are mutually exclusive event.

28. Three persons work independently to decipher a message in Morse code. The respective probabilities of their deciphering the code are $\frac{1}{4}, \frac{1}{3}$ and $\frac{1}{5}$. What is the probability that the message will be deciphered?

29. A class consists of 80 students, 25 of them are girls and 55 boys, 10 of them are rich and the remaining poor, 20 of them are fair complexioned. What is the probability of selecting a fair complexioned rich girl?

30. A university has to select an examiner from a list of 50 persons 20 of them are women and 30 men, 10 of them knowing Hindi and 40 not, 15 of them being teachers and the remaining 35 not. What is the probability of the university selecting a Hindi knowing women teacher?

31. Given the following probability distribution :

x: 0	1	2	3	4	5	6	7
p(x)	: 0	2λ	2λ	λ	3λ	λ^2	$2\lambda^3$

    $7\lambda^2 + \lambda$

    (i) Find λ, (ii) Evaluate $P(x \geq 5)$ and $P(x < 4)$.

32. A die is tossed thrice. A 'success' is 'getting 1 or 6', on a toss. Find the mean and variance of the number of successes.

33. From a bag containing 2 rupee coins and 3 twenty paise coins, a person is asked to draw two coins at random. Find the value of his expectation.

34. An urn contains a white and b black balls c balls are drawn from the urn. Find the expectation of the number of white balls drawn.

35. The probability that a boy will get a scholarship is 0.80 and a girl will get is 0.90. What is the probability that atleast one of them will get a scholarship?

36. One shell from an aeroplane is enough to destroy a target if it strikes it in the centre but does insignificant damage otherwise. In certain conditions the probability of hitting the target is 0.2. How many shells should be used so that the probability of hitting the target is atleast 0.95. Find also the probability of hitting the target if 6 shells are used.

    Hint : Let n shells be used. Probability of not hitting the target = $1 - 0.2 = 0.8$

    So, the probability that atleast one shell hits the target = $1 - (0.8)^n$

    According to question $1 - (0.8)^n \leq 0.95$

    $\Rightarrow 0.05 \geq (0.8)^n$

    $\Rightarrow \log(0.05) \geq n \log(0.8)$

    $\Rightarrow -1.3010 \geq n\,[-0.0969]$

    $\Rightarrow n \geq \dfrac{1.3010}{0.0969} \Rightarrow n \geq (13.42621)$

    $\Rightarrow n \geq 14]$

37. Comment the statement "The mean of Binomial distribution is 3 and variance is 4".

38. A die is thrown 3 times. If getting a 'six' is considered a success, find the probability of (i) 3 successes, (ii) atleast two successes.